D0931488

THE EMERGING UNIVERSE
Essays on Contemporary Astronomy

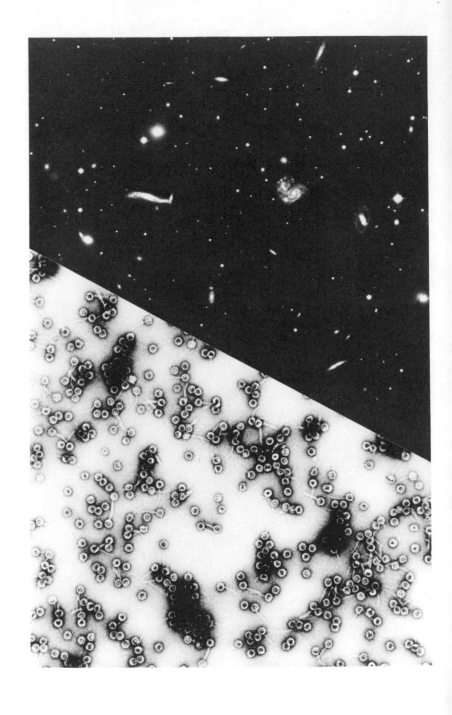

The Emerging Universe

Essays on Contemporary Astronomy

Edited by William C. Saslaw, *University of Virginia,*

National Radio Astronomy Observatory, and University of Cambridge,

and Kenneth C. Jacobs, *University of Virginia*

University Press of Virginia
Charlottesville

THE UNIVERSITY PRESS OF VIRGINIA
Copyright ⊚ 1972 by the Rector and Visitors
of the University of Virginia

First published 1972

Frontispiece. Top: Part of the Hercules cluster of
elliptical, spiral, and distorted galaxies at a distance
of many million light-years. (Hale Observatories
photograph) *Bottom*: A submicroscopic collection
of λ-coliphages, organisms that live and propagate
only by invading much larger bacterial cells.
(Electron-micrograph courtesy of Robert Huskey)

ISBN: 0-8139-0397-1
Library of Congress Catalog Card Number: 72-188526
Printed in the United States of America

Preface

This collection of essays describes our emerging picture of the Universe. Each essay summarizes past knowledge and introduces an important, usually unsolved, problem in astronomy and astrophysics. Some new, untested ideas are presented to illustrate plausible but uncertain developments, and some critical comments are made to warn us against strong belief in those simple models which are so appealing, but unproved. Taken together, these essays provide some insight into the types of questions that astronomers ask, the variety of ways they go about trying to understand the Universe, and some of the results of their search.

This book originated from a series of informal evening lectures on modern astronomy, presented at the University of Virginia during the spring of 1971, and it retains some of their informality. It is a pleasure to thank those persons who made possible this "Cosmology Spring" at Virginia. Laurence Fredrick, director of the Astronomy Department, not only encouraged the series but also obtained the financial support that permitted our distinguished guests to spend several weeks here. These visits were partially supported by the Center for Advanced Studies, and we gratefully acknowledge our debt to its director, Dexter Whitehead. For permission to quote from Bertrand Russell's *The ABC of Relativity*, we are grateful to George Allen & Unwin Ltd.

No expression of appreciation can recompense the exacting labors of Miss Mary King, who typed the manuscript, and Miss Peggy Weems, who drew the diagrams. Of course, none of this would have been possible without the splendid cooperation of the many astronomer-authors.

William C. Saslaw
Kenneth C. Jacobs

Contents

Preface   v

AN INTRODUCTION TO THE EMERGING UNIVERSE
William C. Saslaw   3

1. WHY DOES THE SUN SHINE?
   Leon Mestel   13

2. PLANETARY SYSTEMS
   Shiv S. Kumar   25

3. THE CRAB NEBULA AND PULSAR
   Virginia Trimble   35

4. CHEMISTRY BETWEEN THE STARS
   Lewis E. Snyder   54

5. LIFE IN THE UNIVERSE
   Franz D. Kahn   71

6. GALAXIES: LANDMARKS OF THE UNIVERSE
   Morton S. Roberts   90

7. WHAT OLBERS MIGHT HAVE SAID
   Peter T. Landsberg and David A. Evans   107

8. BIG-BANG COSMOLOGY: THE EVOLUTION
   OF THE UNIVERSE
   George B. Field   143

9. A QUANTUM UNIVERSE: THE BEGINNING OF TIME
   Kenneth C. Jacobs   153

10. WAS THERE REALLY A BIG BANG?
    Geoffrey Burbidge   165

Bibliographical Commentary   181

Index   193

Figures

	Hercules cluster of galaxies; λ-coliphages	*frontispiece*
1.	Earth from Gemini 4	44
2.	The lunar crater Eratosthenes	45
3.	Earth seen from the Moon	46
4.	Mars as photographed by Mariner 7	47
5.	The giant planet Jupiter	48
6.	The ringed planet Saturn	49
7.	A supernova remnant, the Crab nebula	50
8.	Details of the Crab nebula and pulsar	51
9.	Eta Carina nebula	52
10.	Filamentary nebula in Cygnus	53
11.	A Milky Way star cloud	131
12.	The giant spiral galaxy in Andromeda	132
13.	A distant spiral galaxy, NGC 6946	133
14.	The Sombrero galaxy, NGC 4594	134
15.	Explosion of a galaxy, M 82	134
16.	First quasar discovered, 3C 273	135
17.	Disk of our Sun	136
18.	A solar prominence	138
19.	A ball of stars, the globular cluster M 13	138
20.	A young open cluster, the Pleiades	139
21.	Four normal galaxies	140
22.	Four peculiar galaxies	141
23.	Center of the Leo group of galaxies	142

Diagrams

1.1	A Hertzsprung-Russell diagram	14
1.2	Color-magnitude plot for globular cluster M 3	15
1.3	Composite color-magnitude plot for star clusters	16
4.1	Model for a molecular cloud	67

5.1   Atomic binding curve   75
5.2   Stably bound nucleons in nucleus   77
5.3   Unstably bound nucleons in nucleus   78
5.4   Outline of stellar evolution   81
5.5   The A-T base-pair in DNA   84
5.6   The C-G base-pair in DNA   85
5.7   A hydrogen bond   86
6.1   Hubble's galaxy classification   94
7.1   Spherical shell of stars   111
7.2   Classical Doppler effect   118
7.3   Aberration of velocities   121
7.4   Aberration dilution of radiant energy   122
7.5   Contraction of a receding rod   124
9.1   Wheeler's superspace arena   159
9.2   Evolution of type IX cosmology   161

Tables

2.1   Data on the "dark" companions   29
4.1   The twenty-four interstellar molecules reported as of
      January 1972   61
7.1   Models of Universe   129
10.1 Limitations set by source counts on
      various cosmological models   175

THE EMERGING UNIVERSE

Essays on Contemporary Astronomy

An Introduction to the Emerging Universe

William C. Saslaw

University of Virginia, National Radio Astronomy Observatory, and University of Cambridge

Today astronomy is going through an accelerating renaissance. Every two or three years in the last decade there have been new discoveries so surprising and so fundamental that they have reorganized our basic concepts of the Universe. This reorganization is almost as profound as the difference between the concepts of Ptolemy, whose Earth was the center of his Universe, and those of Copernicus, who reduced the Earth to a satellite of the Sun. This reduction was, of course, not easy; for by diminishing the Earth, Copernicus also diminished our own importance to the Universe. Toward the end of this introduction, we will see that there is today a hint, just a hint, that something similar may happen again in the next few years.

In their enthusiasm, astronomers will often tell one another that there is hardly any branch of physical science which does not contribute to their subject. If they are especially enthusiastic, they will sometimes put it another way: cosmology contains all of science. (Psychologists have been known to reply that their subject contains all the cosmologists.) Indeed, we have learned that in order to understand the very large we need to know the very small, and to discover why part of the Universe is alive. The frontispiece of this volume shows the contrast nicely. Above the diagonal slash lies a cluster of interacting galaxies, as seen by the largest telescopes when pointed toward the constellation Hercules. Each of these galaxies contains many billions of stars like our Sun. Below the diagonal, we look through the electron microscope at a collection of molecules which is a million, million, million, million, million times smaller. But these are not just any molecules, for they have arranged themselves into phage λ, a virus which grows in the *Escherichia coli* bacteria found in the human intestine. The icosahedral head of this virus injects DNA through its tail into a bacterium. Inside the bacterium the virus multiplies, and eventually the *E. coli* bursts open to release hundreds of new virus particles.

The heavy atoms that make up these viruses come from the deep interiors of the stars. They were made under conditions of enormous temperature and pressure and were exploded far into interstellar space when a star became a supernova. You and I, this book you hold, and almost everything we see around us originated in such an explosion. Billions of years have passed since this bond was forged between us and the stars. Now we are just beginning to understand it.

Like any other science, astronomy advances through the comparison of observation with theory and the friendly rivalry between observers and theorists. From the tension between new theories and old observations, a new and changing picture of the Universe is emerging. It is this picture—and some of the problems and questions it raises—that is the underlying theme of these essays.

To put the subject in perspective, let us start with the closest part of the Universe and gradually reach outward to survey the heavens. The most familiar regions of our Universe are those nearest to us in space and time. These objects nearest to us have a remarkable property: they preserve their form for fairly long periods of time. For example, the walls of a room do not pulsate in and out, and a chair does not suddenly rise up to the roof. This behavior of matter is peculiar to cold, stable atoms and molecules bound together by nuclear and electromagnetic forces. But these rather dull conditions, although necessary for our kind of life, do not represent the nature of the Universe at all well. Probably less than about one atom in 10,000 has the simple order we see around us.

Evidently, to gain some perspective we must first rise above our surroundings. In imagination, this is quite easily done, and a view of Earth from Gemini 4 (Fig. 1) serves to stimulate the imagination. If it is a clear day as we look toward the setting Sun, we might pause to wonder at this origin of vast quantities of energy flooding indiscriminately into space. An atom near the center of the Sun has a temperature of 10 million degrees and a pressure one billion times the pressure we feel on Earth. Under these conditions no rigid structure can exist, and the atoms behave very differently from their quiet counterparts on Earth. Why these atoms cause the Sun to shine, and how long it will continue, are the subjects of Leon Mestel's essay. The Sun contains more than 100,000 times the amount of matter of our planet. Already, we see what a rare thing we are in the Universe.

But we have just begun. After the sunset, we see our satellite neighbor, the Moon. Its surface: dust and cratered rock (Fig. 2). Its only form of life: the wandering astronaut. Here again is one of those minor places where atoms are arranged and ordered. Looking back from the Moon we have a remarkable view of our planet, the Earth, rising above the Moon's horizon (Fig. 3). But we must not pause here because there is the Universe to explore. And so, outward through the Solar System: past Mars (Fig. 4), so cold that its polar caps are not snow but probably crystals of frozen carbon dioxide or dry ice. Past the giant planet Jupiter (Fig. 5), and on beyond Saturn (Fig. 6), nearly ten times as far from us as we are from the Sun.

As we go still farther, so that the remaining planets of our Solar System lie behind us, something new happens. There is a fundamental change in the quality of the information we receive with the naked eye or with our telescopes, for now we see mostly mere points of light. The stars are so distant that their surfaces cannot be resolved. In fact the nearest star is 100,000 times farther from Earth than Saturn is. Our rate of reaching outward is rapidly accelerating. Still, we can get much information about the stars from their color and the spectrum or frequency distribution of their radiation. The spectrum tells us the kinds of atoms radiating in a star's atmosphere, and also how fast the star is moving.

These stars are so bright that we are unable to see any dimly illuminated planets they may have. Only if the planet is so massive that it perturbs the star's gravitational orbit might we hope to notice it. This disturbance may be detected by photographing the star's position for many years and looking for small wobbles in its motion on the sky. Shiv Kumar discusses whether planets exist in other solar systems and whether they too might support life.

Surrounding us are billions of stars. Leon Mestel shows us that as these stars live, so also must they consume their own substance and eventually expire. The less massive stars are now in their primes with many eons of existence before them, while newborn stars are even now appearing from the interstellar gas and dust. But what about those massive stars formed long ago, which burned their fuel so prodigally that they should now be dead—where are they? Many have undoubtedly become small white (or even black) dwarfs, slowly cooling to invisibility, but about once every 100 years in our Galaxy a more spectacular form of stellar demise occurs. Virginia Trimble describes the result of such a supernova explosion, and the formation of the pulsars, in her essay on the Crab nebula (Figs. 7, 8).

Between the stars there is not-quite-empty space. Clouds of gas containing masses hundreds, sometimes thousands, of times greater than the mass of the Sun populate these interstellar regions (Figs.

9, 10). Most of this gas is hydrogen; about 25 percent (by mass) is helium. There is also a small but very remarkable fraction of heavy atoms. Surprisingly, these heavy elements are often found to cluster in complex molecules. Lewis Snyder describes the discovery and the nature of these molecules in his essay, and Franz Kahn suggests in his essay on life in the Universe that they could be the precursors of life here on Earth.

The stars we see in most parts of the sky are fairly sparsely distributed, but there are some regions clouded by a dense mist of stars (Fig. 11). These, of course, are the concentrated parts of our own Galaxy—the Milky Way. The Sun is only one of nearly a million, million suns all revolving together in a great complex of stars and gas, a complex so huge that only fifty years ago it was thought to be the full extent of the Universe. But today we know that it, too, is but a small part of a larger system.

If we look between the stars of our own Galaxy, we see small patches of hazy light. When the largest telescopes are turned toward these nebulae they reveal other galaxies, often as large, sometimes larger, than our own. The closest large one is in the constellation Andromeda (Fig. 12), almost 2 million light-years away. Because light travels with a finite, constant speed of 186,000 miles per second, the light we see now left the Andromeda galaxy when man was just beginning to evolve on this planet.

But there are galaxies farther away than Andromeda, for example, the spiral galaxy known as NGC 6946 (Fig. 13) and the famous Sombrero galaxy (Fig. 14) in the constellation Virgo. Not all galaxies lead such regular quiescent lives. The galaxy M 82 (Fig. 15) has exploded recently (by astronomical standards of age), and the symptoms are shown in the red light of hot hydrogen gas. We are beginning to suspect that such outbursts may be part of the evolution of many galaxies.

Galaxies provide our main clues to the development of the Universe. They are close enough to study in detail, yet large enough and far enough apart that the history of the Universe has influenced their structure and distribution. Morton Roberts describes the progress we are gradually making in deciphering these Rosetta stones of astronomy.

When the distant galaxies were first examined in the early decades of this century, it was discovered that most of them are moving away from us. Moreover, they are also moving away from each other. This led to the idea that the entire Universe is expanding. All parts are uniformly moving away from each other, like raisins in a rising cake, as has often been said. The discovery of the expansion of the Universe was one of the most significant developments of early twentieth-century science. It forms the basis of modern cosmology. The four concluding essays by Peter Landsberg and David Evans, George Field, Kenneth Jacobs, and Geoffrey Burbidge describe aspects of cosmology from rather different points of view.

This basic fact—the expansion of the Universe—is found by looking at the distribution of light in the spectra of distant galaxies. Every kind of atom, such as carbon, oxygen, or iron, radiates light at certain particular frequencies or colors. If the galaxy is moving away from us, we see this radiation, but it is shifted toward the red end of the spectrum—the famous red shift. The faster the galaxies recede, the greater is their red shift.

The galaxies we have just been looking at are still close enough to see in detail with the largest telescopes. If we look even farther into space, into regions almost inconceivably remote, we find galaxies so far away that we can see them in groups. In the Hercules cluster (frontispiece), we find an unusual display of many types of galaxies. Here are spirals, double spirals, ellipticals, and unusually elongated galaxies probably distorted by the tidal action, the gravitational pull, of their neighbors.

Looking toward yet more distant clusters, it becomes hard to distinguish the galaxies from the nearby stars. Both look like points of light. But the galaxies are just a little bit fuzzy or elongated, thus revealing their identity.

And even farther out in space—but wait! Let us pause in our journey, lest by hurrying too fast we cease to understand what we see. For now a discontinuity occurs in our understanding. The picture presented so far is more or less where things stood ten years ago. It seemed as though all the major constituents of the Universe had been observed. The goal was to see farther by finding more and more distant galaxies and clusters of galaxies. By careful observations of these, at both optical and radio frequencies, it seemed possible to determine the rate at which the Universe expands, and which of the models from Albert Einstein's general theory of relativity best fits our actual Universe. All this began to look fairly straightforward.

But we had a warning from the past, from a very distinguished astronomer, John Herschel. In January 1820 a group of eight gentlemen met in a tavern in London to found the Royal Astronomical

Society. Herschel was among them, and after the meeting he prepared a summary of the motives and objects of the society. There was one sentence present in the original draft of his summary but not in the final one. He wrote: "Yet it is possible that some bodies, of a nature altogether new, and whose discovery may tend in future to disclose the most important secrets in the system of the universe, may be concealed under the appearance of very minute single stars no way distinguishable from others of less interesting character, but by the test of careful and often repeated observations." We do not know what caused Herschel to delete this percipient remark from the final version. But it is clear that he was not letting mere appearance confine his imagination. And almost 150 years later he turned out to be right. It was a long time to wait!

What happened was this. In the late 1950's and early 1960's a number of new radio sources were discovered at Cambridge, England. Until then many radio sources had been identified with peculiar optical objects at the corresponding position in the sky. But the only objects close to these new radio sources were ordinary-looking stars, ordinary-looking, that is, until studied with the spectrograph at the largest telescope, the 200-inch at Mt. Palomar in California. When the spectrum of the star closest to one of these radio sources was analyzed, it made no sense. No star with such a spectrum had been known before. Interpreting these results was the great puzzle. Finally this mystery was broken in 1963 when M. Schmidt recognized the lines in the spectrum as belonging to red-shifted hydrogen. But it was hydrogen red shifted by such a large amount that if the cosmological red shift arguments are correct, this starlike object is one of the most distant things in the Universe, much more distant than the clusters of galaxies we just mentioned. The object is known as 3C 273, and its image is in Figure 16. It was called a *quasi-stellar radio source*, later abbreviated to *quasar*.

This quasar is so far away that when the light that reached us several years ago to make this photograph left the quasar, there was no life on Earth. And yet, this is one of the closest of the quasars. It has a red shift of 0.16. Today, nine years after 3C 273 was uncovered, about 200 quasars are known. Many of them have red shifts greater than 2, some as much as 2.9. If this red shift is due to the expansion of the Universe, it means we are seeing objects in the Universe as they were almost 8 billion years ago. We are looking through 80 percent of the history of the Universe since it started expanding, according to the simplest models of Einstein's general theory of relativity.

Because these objects are so distant in space and time, they have other remarkable properties as well. Many of them appear quite bright, as bright as many stars in our own Galaxy, the Milky Way. To appear this bright at their great distances, the quasars must produce enormous amounts of energy. A single typical quasar produces as much energy as 10,000 galaxies like our own. But the story does not end here. Quasars are not as large as our own Galaxy. In fact most of them are quite tiny by comparison. If quasars were placed side by side in all directions within our Galaxy, it would take about 100,000 quasars to fill it. And if our Galaxy were filled with quasars, it would be a billion times brighter than it is, and there would be no such thing as a night sky. These astonishing conclusions about the distance, the brightness, and the small size of quasars have caused some astronomers to question the principle on which the conclusions are based, namely, the cosmological red shift. Although several alternative explanations have been suggested, none has yet proved as plausible as the cosmological red shift. But this is still an open and exciting subject for debate, especially since the spectra of some quasars can be explained only if they have more than one red shift.

The quasars were the first of the new things in heaven to be discovered in the last decade. We still do not know what they are or the secret of their vast store of energy. But we have some ideas. According to one idea, we are looking at condensed galaxies with large numbers of stars so close together that in the nucleus of the galaxy stars collide with each other and explode. In another theory, matter is releasing great bursts of light from an even more condensed object near the center which forms a general relativistic singularity. The radiation is connected with this singularity. In yet a third theory, we are seeing the birth of a galaxy in a huge flash of luminous star formation. To decide between these and other ideas, we need more observations and more large telescopes with which to make them.

When we see the quasars sending signals from the early Universe, we begin to wonder whether there may be any other way of observing the distant past. The quasars were the first new type of object discovered recently. But also, about six years ago, astronomers found that the Universe was filled with a new kind of electromagnetic radiation. This radiation does not come from stars, or galaxies, or even from quasars. Rather, we now think it comes from a past even more distant than the quasars. We are not quite sure yet, but the radiation appears to have a temperature of about 3 degrees on the

absolute scale. On this scale of temperature, a room has a temperature of about 300 degrees. So the new radiation is rather cold, but
even so this glimmer of warmth uniformly fills the entire Universe.

There is one theory of cosmology, based on Einstein's equations
and studied especially by the late George Gamow, that predicted
this radiation twenty years ago. This is the big-bang cosmology. In
this theory, the Universe was once filled with very dense and very hot
matter and radiation, with densities and temperatures billions of
times greater than any we experience on Earth. From this primordial
fireball, the Universe expanded, gradually cooling; and as it cooled
below 3000 degrees, small irregularities in the gas began to grow.
They were a little more dense than their surroundings and managed
to hold themselves together by gravity as the Universe expanded.
Eventually these condensations began to contract and some, while
contracting, happened to merge together. In this way the galaxies
and clusters of galaxies began to form.

Within each forming galaxy, another new activity was also beginning. Smaller clouds of gas, pulled together by gravity, were contracting faster than the overall galaxy. As these small clouds contracted, their temperature began to rise. And when their temperature
reached several million degrees, atoms of hydrogen crashed into
each other with such force that they fused together to become deu-
terium and helium. In the fusion process energy was released, and
these thermonuclear reactions became self-sustaining. The small
clouds of gas had become stars, and the galaxy began to shine.

Throughout all this process, some of the radiation from the early
dense phase of the Universe persisted. Ignored now by the matter,
it continued to fill the Universe, cooling as the Universe expanded.
It is this radiation, cooled to 3 degrees, that many astronomers
believe we are seeing today.

The presence of this radiation has been taken by most astronomers
as the main evidence contradicting the opposing steady-state cosmology. In the steady-state theory, the Universe never passed
through a dense stage. Rather it always had the same general appearance we see now. In order for the Universe to keep its present
appearance while expanding, matter must be created, from nothing,
so that the density does not drop. No one has yet found a natural
explanation for the 3 degree background radiation in such a universe.

Einstein's theory of relativity, which describes the big-bang cosmology, also predicts another phenomenon. This phenomenon may

be the latest major discovery in astronomy. Einstein's equations predict a new type of radiation—gravitational radiation. This radiation is a form of energy emitted by rapidly moving objects such as atoms, stars, or even galaxies whose acceleration changes rapidly in time. Because this radiation is usually very weak, it is difficult to detect. But recently a group led by J. Weber at the University of Maryland has built a detector for gravitational waves. As they increased the sensitivity of the detector, they began to find large numbers of strange signals.

As the instrumentation was improved these signals continued to persist. Currently the most reasonable interpretation is that the signals are indeed pulses of gravitational radiation. They appear to come from a very special direction in the sky: the nucleus of our own Galaxy. There are still some doubts about this detection, for technical reasons. But because of its importance, at least eight laboratories are now trying to repeat the experiment. In several years, we should be much more sure of the answer.

If these observations are correct, then something very interesting is happening. The gravitational radiation antenna now detects a few strong pulses every week. This means that sharp explosions are taking place at this rate in the center of our Galaxy. We can calculate the amount of matter that is changed into gravitational radiation and lost from the Galaxy in each explosion. It turns out to be about 100 times the mass of the Sun each year.

One hundred solar masses per year may not sound like very much compared to the 100 billion suns in our Galaxy. But if this rate of mass loss has continued, year after year, for the entire history of the Galaxy, we are led to a rather startling result: when the Galaxy formed, it was at least 20 times more massive than it is now! In other words, most of our original Galaxy turned into gravitational radiation and fled into far-off regions of space. Our Galaxy is a mere shadow of its former self. If this is true of most galaxies, then all the Universe we see through our telescopes represents only a tiny fraction of the total matter and energy that there is.

There are two ways to explain away this conclusion. One is to suppose that our Galaxy just started radiating recently, say a billion years ago. The problem with this solution is that the a priori chance of it being correct is only about 10 percent. The second way out is that the observations may be wrong. That is why so many people are interested in repeating them using somewhat different techniques. If, after further investigation, Weber's results are confirmed, we will have to understand the Universe anew. The role of our Galaxy, and Sun, and planet will be even further diminished in this emerging view of the Universe.

Perhaps to complete this introduction, we should return from our exploration of the distant Universe, return to our immediate surroundings. Here we can put our mental journey itself in perspective by recalling Bertrand Russell's comment:

To those who have lived entirely amid terrestrial events and who have given little thought to what is distant in space and time, there is at first something bewildering and oppressive, and perhaps even paralyzing, in the realization of the minuteness of man and all his concerns in comparison with astronomical abysses. But this effect is not rational and should not be lasting. There is no reason to worship mere size. We do not necessarily respect a fat man more than a thin man. Sir Isaac Newton was very much smaller than a hippopotamus, but we do not on that account value him less than the larger beast. The size of a man's mind—if such a phrase is permissible—is not to be measured by the size of a man's body. It is to be measured, in so far as it can be measured, by the size and complexity of the universe he grasps in thought and imagination. The mind of the astronomer can grow, and should grow, step by step with the universe of which he is aware. And when I say that his mind should grow, I mean his total mind, not only its intellectual aspect. Will and feeling should keep pace with thought if man is to grow as his knowledge grows. . . .

We are beset in our daily lives by fret and worry and frustration. We find ourselves too readily pinned down to thoughts of what seems obstructive in our immediate environment. But it is possible, and authentic wise men have proved that it is possible, to live in so large a world that the vexations of daily life come to feel trivial and that the purposes which stir our deeper emotions take on something of the immensity of our cosmic contemplations. Some can achieve this in a greater degree, some only in a lesser, but all who care to do so can achieve this in some degree and, in so far as they succeed in this, they will win a kind of peace which will leave activity unimpeded but not turbulent.

1. Why Does the Sun Shine?

Leon Mestel
University of Manchester

When asked to study the Sun, as opposed to stars in general, the theoretical astrophysicist suffers from an *embarras de richesses*, for we are close enough to the Sun for our observational colleagues to have supplied us with an awful lot of detail to keep us busy. Just look at a photograph of the solar surface: already one sees the large sunspot groups and local fine structure (Fig. 17). More recent observations, taken from balloons, give much more detail of this surface granulation, caused by small-scale motions. The sunspot regions are found to be permeated by strong local magnetic fields. Sunspot motions allow us to infer that the solar surface is not rotating as a rigid body; there is a definite increase in rotational velocity toward the solar equator. From observations during solar eclipse we find that the Sun is surrounded by a hot corona, which we now know is expanding; we call this the *solar wind*. Gigantic archlike prominences are observed (Fig. 18). Violent eruptions—solar flares—cause a sudden increase in the force of the solar wind and consequent radio disturbances on Earth.

However, now we ignore all these fascinating details and restrict our study to the more fundamental question, What is it that makes the Sun shine? But we can be sure that the answer will apply also to the millions of other stars in our own and other galaxies. We have learned enough from the Copernican Revolution not to regard ourselves as in any way privileged. I am sure that plenty of other stars similar to the Sun have planetary systems and have supplied the warmth needed to enable other men also to develop to the point where they are capable of destroying themselves. Just to emphasize that we are not alone, look at a densely populated part of the Milky Way (Fig. 11), the disk that contains the bulk of the stars in our own Galaxy. Then there is our neighbor and big brother—or sister—the great spiral galaxy in the direction of the constellation Andromeda (Fig. 12). Next, look at Messier 13 (Fig. 19), a typical globular cluster, an old subsystem of our Galaxy, distinguished not only by its highly symmetrical shape and the large number of stellar components but by the fact that its brightest stars are red. In contrast to Messier 13 is the young galactic cluster, the Pleiades (Fig. 20), believed to be the seat of new star formation, where the brightest stars are blue. Any convincing physical theory of the solar luminosity must be comprehensive enough to explain the basic properties of these many different types of stars and stellar systems.

Let me now introduce a famous diagram, associated with the astronomers Hertzsprung and Russell (Diagram 1.1). Plotted vertically is the intrinsic luminosity of the star considered; horizontally is the temperature of the surface—a measure of the color of

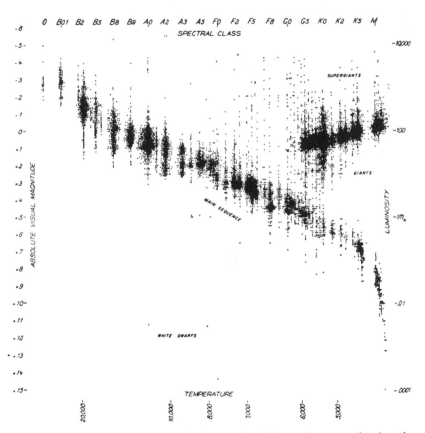

Diagram 1.1. A Hertzsprung-Russell diagram for 7000 nearby stars, showing the spectral class of each star (scale at *top*), which correlates with the stellar temperature (*bottom* scale), and each star's magnitude (*left-hand* scale), which is a logarithmic measure of the stellar luminosity (*right-hand* scale). (After W. Gyllenberg, Lund Observatory)

the light emitted, which is closely correlated with the spectral lines absorbed by hot atoms near the surface. Just to make things difficult, the temperature increases to the left—red stars are on the right, white and blue stars on the left—and although luminosity increases upwards, astronomers usually talk about magnitudes, a quantity that decreases with increasing luminosity. The most obvious feature is that the diagram is not full: notice in particular the main sequence, the red giants, and the faint white dwarfs. This particular diagram is a composite of stars from different parts of the galactic field. More significant for our theoretical studies are the H-R diagrams for individual *clusters*, gravitationally coupled groups of stars that probably all formed within a comparatively short time. In the plot for the globular cluster M 3 (Diagram 1.2), notice the well-defined turnoff point of the giant branch from the main sequence. A famous composite diagram produced by A. Sandage describes schematically the

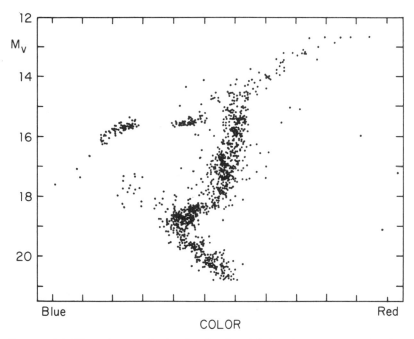

Diagram 1.2. The color-magnitude plot for the globular cluster M 3. Color is a measure of stellar temperature; blue stars are hotter than red. All of these stars belong to one star cluster. Note the lower main sequence (*bottom*), the turnoff to the giant region (*right*), and the horizontal branch (*upper left*). (Reprinted by permission from H. L. Johnson and A. R. Sandage, *Astrophysical Journal* 124: 379, 1956 © 1956 by The University of Chicago).

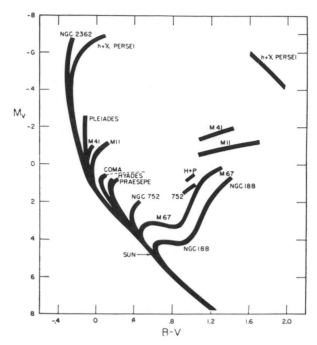

Diagram 1.3. A composite color-magnitude plot for several star clusters. Here *B-V* represents color, with hot blue stars to the left-hand side. The lower the main-sequence turnoff point, the older is the cluster. (From A. R. Sandage, *La Plata Symposium on Stellar Evolution*, 1962)

main characteristics of different clusters (Diagram 1.3). Now let us turn to theory.

In the last century the question asked by physicists tended to be not Why does the Sun shine? but the superficially similar question, What is the supply of energy upon which the Sun draws? An answer to this query was given by Kelvin and Helmholtz, who argued that the Sun lives on its gravitational capital. A boulder held at the top of a cliff has potential energy by virtue of its position in the Earth's gravitational field; but when it is dropped into the valley its motion toward the center of the Earth corresponds to a conversion of potential energy into energy of motion, or kinetic energy. By analogy, if a gas sphere contracts, then each particle moves closer to the center of the sphere's gravitational field and so liberates potential energy. We know how much energy the Sun radiates per second; we can therefore estimate at what rate the Sun's radius must shrink to supply the solar luminosity, and so how long the Sun would take to reach its present size from an earlier state, *e.g.*, from the time it had effectively condensed from the interstellar medium. The answer they found was embarrassingly short—about 30 million years—as compared with contemporary geological estimates. Significantly, the answer was not so very different from Kelvin's estimate for the Earth's age based on its rate of cooling. We now know that both estimates are wrong for rather similar reasons; they ignore the supply of energy from nuclear sources. In the sun, *thermonuclear fusion* of light nuclei into heavier nuclei releases heat. In the Earth, energy is released by *radioactive decay*, the spontaneous fission of heavy elements into lighter ones. But even if the early estimates had been correct, one would have remained dissatisfied with having to assume the value of the solar luminosity, instead of predicting it. Further, painstaking observations have revealed that many stars satisfy a mass-luminosity relation, the luminosity increasing as a strong power of the mass. The answer to the question Why does the Sun shine? must surely be capable of explaining this relation as well as accounting for at least the broad features of the H-R diagram.

One is tempted to argue: The Sun shines just because it has thermonuclear sources that keep on pumping energy in, and the Sun must shed this energy. But suppose the thermonuclear sources at the solar center are by fiat cut off; what would happen? Let us first frame a similar question for the Earth, and imagine the radioactivity within the Earth removed. Heat continues to flow out by conduction from the hot interior to the cooler surface, to be radiated away, and the Earth cools; indeed, it was by assuming that the Earth lives on its thermal capital that Kelvin estimated its age. Now the temperature of the Earth is of comparatively minor importance for its structure. The solid crust, the liquid core, and the solid inner core all have enough strength to withstand the crushing force exerted by the Earth's self-gravitation: the Earth can cool to zero temperature without running into serious difficulties! But the Sun's gravitation is so much greater that only gases are capable of exerting the pressures required to keep the Sun in equilibrium. (We can be sure that this equilibrium must be very closely maintained; if there were a departure from equilibrium of no more than 1 percent, we would see the Sun's radius alter markedly in about half a day.) We can take over from high-school physics the law that a gas far from condensation exerts a pressure proportional to its density and temperature; this is the so-called *perfect gas law*. The pressure is a manifestation of the random motions of the particles of the gas, and these motions decrease with decreasing temperature. Thus, for the star to be able to withstand its own gravitational field, its internal temperature must be sufficiently high; and the smaller the radius, the more powerful is the gravitational pull, and correspondingly, the higher is the internal temperature. Within the Sun at its present radius the mean temperature is about a million degrees, and the central temperature about ten million, as compared with a surface temperature of about five thousand degrees. We emphasize that, in contrast to the case of the Earth, these temperatures are essential for the very existence of the star as a body in mechanical equilibrium.

We have our first check on the consistency of our model here. A simple calculation shows that the average density of the Sun is similar to that of liquid water; yet we are treating the solar material as if it behaved like a perfect gas. There is no reason for incredulity. A substance behaves like a gas if its particles can move freely for distances large in comparison with the average interparticle distance before being seriously deflected, or equivalently, if the random kinetic energy is large compared with the mean energy of the electrical forces between the particles. At the temperatures that we have found within the Sun, this condition is amply satisfied.

Now we know that when one part of a body is hotter than another, heat flows in such a way as to try to iron out the temperature differences. So there must be a leak of energy from within the star to its surface; and this must continue, because the temperature gradient is maintained by the gravitational force. Thus the star radiates because it cannot help doing so. The presence or absence of thermonuclear sources is now seen to be almost an irrelevancy. If there are no sources, then the star behaves as discussed by Kelvin and Helmholtz: It contracts steadily, and the gravitational potential energy released partially makes up for the energy radiated away and partially heats the star to the higher temperature that is required for equilibrium at the smaller radius. But, unlike the Earth and human beings, which cool as they lose heat and so are able to approach the temperature of their surroundings, a star like the Sun gets hotter as it loses heat. The apparent paradox is similar to the behavior of a satellite orbiting the Earth, with the gravitational pull of the Earth balanced by centrifugal force. The frictional drag of the Earth's atmosphere takes energy away from the satellite, which moves gently into an orbit closer to the Earth with a kinetic energy greater than before, so that centrifugal force can still balance the increased gravitational pull of the Earth. The satellite's kinetic energy is the analog of the heat energy of the star; the frictional energy loss corresponds to the star's radiation; and the gravitational energy decrease, as the orbit shrinks, supplies both the frictional loss and the increase in kinetic energy.

Let us now look more closely at the physical conditions within the Sun and stars. At the enormous temperatures we have found, atoms of hydrogen and helium are stripped of their electrons, and the heavier elements present also lose some electrons, so that the gas is a hot *plasma*, a mixture of free electrons and positively charged ions. Such a gas is an excellent emitter and absorber of electromagnetic radiation. At every point within the star there is built up a density of electromagnetic waves, or of *photons* in quantum terminology, with mean wavelength fixed by the local temperature; the waves are mainly X-rays deep within the star but optical radiation near the surface. An individual photon emitted by the gas will travel about a centimeter before it is reabsorbed or scattered. If the temperature were uniform, then the number of photons traveling in all directions would be the same. As it is, there are more photons per unit volume in the deeper hotter regions than in the cooler outer regions, and therefore there is a slight net flow of radiation from the center outward. In stars this flow by *radiative transfer* is far more efficient than the flow by conduction (the normal process of heat transfer in solids). From our knowledge of the opaqueness of the stellar material, we can calculate the radiative flow from the center to the surface, and we identify this with the luminosity of the star. The transparent surface layers acquire the right temperature to radiate into space the energy received from the deep interior. Theoretically, the luminosity is found to vary strongly with the mass and composition of the star but to increase only slowly as a star of given mass contracts.

As a nascent star contracts from the interstellar medium, it begins to grow hot. At a definite central temperature and radius it reaches the point where the liberation of thermonuclear energy (by the transmutation of hydrogen into helium) at its center balances the energy leaking from its surface. The contraction is halted, and the star settles down into the state in which it spends most of its active life. Depending upon its mass, the star is now somewhere on the so-called *main sequence* (see Diagram 1.1). The main sequence is the curve of more or less homogeneous stars of different mass, all generating from nuclear sources the energy balancing their spontaneous heat loss. The mass-luminosity relation also emerges from the theory.

Inevitably, things are more complicated than I have indicated, and one fails in one's duty if this is not at least hinted at. Sometimes the material of a star becomes unstable and boils, so that radiative transfer is augmented by *convection* currents, as in a kettle or in a windy atmosphere. This occurs most strikingly in the earlier pre-main-sequence phase; the corresponding evolutionary path in the H-R diagram is now more nearly vertical than horizontal, with the luminosity depending strongly on the radius but decreasing with the star's contraction. The unstable T Tauri stars are believed to be located at the end of this contraction phase.

The next task of the theorist is to follow the evolution of a star as nuclear transformations change the composition and structure, and in particular to explain the H-R diagrams of the different star clusters. Much progress has been made here, and in particular we have learned how to estimate the age of a cluster from the point where the sequence of giant stars branches off from the main sequence. There are even clusters whose H-R diagram cannot be understood without our postulating that the less massive stars have not yet had time to contract to the temperatures where nuclear burning starts: the lower main sequence is missing, being replaced by a curve that slants more to the right. But we must remember that we have not solved but only postponed the problem of how the star will spend its retirement from active life. Nuclear energy liberation by the buildup first of helium and then of carbon and heavier elements from hydrogen cannot proceed beyond the "iron peak." To build up heavier elements requires a supply of energy; as we all know, energy is liberated by the fission of uranium! Once a star has exhausted its nuclear capital, it must resume its overall contraction. It still has no option but to radiate; so it must resume living on its gravitational capital, contracting to higher and higher densities, with its luminosity staying constant or even increasing slightly. We were put the question Why does the Sun shine? and we are finding the answer—Because the very conditions of its existence force it to—an embarrassment. Can we believe the process goes on indefinitely? We are in the predicament of the sorcerer's apprentice; what new magic principle can we introduce that will call an ultimate halt to the process?

In the 1920's the developing quantum theory was having striking success in interpreting the spectral lines emitted by excited atoms in terms of electrons jumping from one quantum orbit to another of lower energy. But there were anomalies: some lines failed to appear.

Wolfgang Pauli clarified the situation by enunciating his famous *exclusion principle*. There is a limit to the number of electrons that can occupy the same quantum orbit. Each electron in an atom requires four numbers to describe it completely, but no two electrons can have the same four numbers. It is this trade-union demarcation rule that removes the anomalies, for the only way that the missing spectral lines could have been emitted is by an electron hopping into an orbit that is already occupied.

Consider a gas of free electrons, as inside a star. Each electron is now described classically by its instantaneous position and by the three components of its momentum. But it is a fundamental of quantum mechanics, enunciated by Heisenberg, that there is a limit to the precision with which we can assign simultaneously both position and momentum to a particle; this is the *uncertainty principle*. Mathematicians like to invite us to stretch our imagination a little and think of a six-dimensional space, three dimensions representing the three directions of our physical space, and the others the three independent directions of the instantaneous momentum of a particle. Now divide this space into boxes, each with the same six-dimensional volume $(h/2\pi)^3$, where h is Planck's constant. The finite size of these boxes is the analog of the set of discrete orbits of electrons within an atom. The Pauli principle allows each box to contain at most two electrons, and these two must have oppositely directed spins. Occupancy by more than two is unlawful, if not positively dangerous.

How does this rule for studying the electron gas affect the problem of stellar structure? At temperatures and densities similar to those in the Sun, the effects are small. There are so many boxes for the electrons to choose from that it is improbable that two will want to occupy the same one; so the classical gas law remains valid. But as the burnt-out star contracts, the electron pressure law changes. The crucial new feature is that the gas exerts a strong pressure even if it is at zero temperature. This is perhaps a strange idea, as we tend to think that temperature is a direct measure of the motion of gas particles. Indeed, the particles of a cooling classical gas would all crowd into boxes corresponding to lower and lower energy of motion. But for the electrons to behave in this way would be an outrageous violation of the Pauli principle. As the gas cools, each electron pair attains the lowest-energy box available to it, but only a privileged few can have really small energies; the mean energy and pressure of this so-called *degenerate gas* remain finite and large.

There is no hope that the electrons will recombine with the ions and so cease to be free particles exerting a pressure, for at these densities the radius of a quantum orbit—itself fixed by Heisenberg's principle —is larger than the mean distance between the ions, which in fact tend to arrange themselves into a crystalline lattice, through which the electrons can move more or less freely. We can thus construct a new type of stellar model, in which the intense gravitational field is balanced by the pressure of the degenerate zero-temperature electron gas. Such a star is known as a *black dwarf*. It cannot radiate energy, as the whole star is in its lowest energy state. It has achieved nirvana as a rather dense body—about a ton to the cubic inch. Because of the Pauli principle, it can withstand gravity without needing to be hot inside, and so without the energy loss forced on a non-degenerate star.

A black dwarf can be detected only by its gravitational effects. However, the zero-temperature models are excellent approximations to the *white dwarf* group of stars in the lower left-hand corner of Diagram 1.1. They have masses similar to the Sun's, but radii similar to the Earth's. If they consisted of classical gas, they would be somewhat brighter than the Sun. In fact, they are several hundred times fainter, but because of the small area of the radiating surface, most of them do not have the dull red color that we tend to associate with faint objects but instead emit a whitish light—whence their name. They are slowly cooling toward the ultimate black dwarf state; in fact the older and fainter white dwarfs have already become so cool that they ought to be called "red white dwarfs"—the term *red dwarf* being preempted for the red stars at the lower end of the main sequence. Just as in Kelvin's Earth model, it is the existence of a pressure that does not depend on temperature that enables a white dwarf to cool as it loses energy.

However, it turns out that only stars of mass less than about 1.2 solar masses can achieve equilibrium as black dwarfs. This is a consequence of the special theory of relativity, which predicts that at the still higher densities within more massive stars the pressure of the degenerate electrons would not be strong enough to withstand gravity. The limited observational evidence on white dwarf masses is consistent with this upper limit; the best-known white dwarf—the faint companion of Sirius—has a mass nearly the same as the Sun's. Now it is certain that during their post-main-sequence evolution stars do lose mass—either gently or violently—but we have no reason to suppose that all stars will oblige us by spewing off enough mass so that the relic is below the 1.2 solar mass limit. Thus the white dwarf is only a partial solution to our dilemma. Burnt-out stars of mass greater than the limit would continue to heat up, radiate, and contract. But when the central density has increased by a largish factor—a hundred million or so—the ions and electrons will have combined to form a gas of *neutrons*. These are uncharged particles, of mass nearly the same as the hydrogen atom, which also satisfy the Pauli principle and so exert a strong zero-temperature pressure, augmented by the strong mutual repulsion of the neutrons. At these densities the changes in the gravitational pull predicted by the general theory of relativity also become important. Our ignorance of the details of the interneutron forces leaves some uncertainty in the predicted maximum mass that can exist as a cold *neutron star*; but the most favorable estimates yield an upper limit of about two solar masses.

Thus we have been led to conceive of bodies far more fantastic than white dwarfs, which are now "old hat." A neutron star is a body of mass comparable with the solar mass, and yet with a radius of no more than ten kilometers! Bodies of such enormous density can rotate very fast without centrifugal forces becoming too strong for gravity—just as the stronger the string holding a ball, the faster we can whirl it before the string snaps. Rotation periods of the order of a second are possible; and it is confidently believed that the periodic radio emission from *pulsars* originates from rotating magnetized neutron stars, the period being identified with the period of rotation.

There is a very short-period pulsar within the Crab nebula—the famous relic of a supernova observed and chronicled by the Chinese about a thousand years ago (Fig. 7; see also Essay 3). This fits in well with current ideas that link the *supernova* phenomenon to the collapse of a massive star until its central regions are of nuclear density. The slowing-up of the collapse of the stellar core by the neutron pressure yields an enormous temperature—about a thousand billion degrees—with the associated thermal pressure high enough to blow off the stellar envelope, leaving us a cooling neutron star, observable as a pulsar, and an expanding supernova remnant.

But suppose the collapsing star is unable to blow off enough mass to settle down as a neutron star. Then the only alternative left for the star is continuing collapse into a *black hole*, the Ultima Thule of gravitational collapse. This is a characteristic prediction of general relativity. The curvature of space-time is so great locally that no light or matter can leave. Any infalling matter loses its identity as it is torn apart by tidal forces. Only the most fundamental quantities—the mass, the electric charge, and the angular momentum—are required to determine the ultimate state; and these quantities can in principle be determined by the motion of test particles in orbits about the black hole. There are high hopes that black holes will be discovered within the next decade.

We can imagine an unfortunate observer attached to the collapsing mass, emitting his last signals—a cry of help—before he crosses the gravity barrier. This signal will be received by a distant observer for all time, because of the effect of the gravitational field on passage of time; the light becomes of steadily decreasing frequency, and one will need first an infra-red and finally a radio-telescope to receive it. But there will be no news from the infernal regions beyond the date (by his reckoning) of the unfortunate victim's death.

The story has reached its end, but I am only too conscious of how much important and challenging detail has been ignored. I have hardly mentioned stellar rotation, and the related questions of the origin of double stars and planetary systems; nor have I discussed variable stars, the strongly magnetic stars, planetary nebulae, and novae. I have skimmed over post-main-sequence stellar evolution and the detailed interpretation of the H-R diagram and the build-up of the heavy elements. We have still very imprecise ideas on the formation of stars from the interstellar gas. So much remains to be done, and life is short. But it is comforting to know that one's livelihood is guaranteed; and perhaps one can derive moral strength from the second-century Rabbi Tarphon, who wrote (admittedly apropos of a rather different way of regarding the Universe): "It is not thy duty to complete the work, but neither art thou free to desist therefrom."

2. Planetary Systems

Shiv. S. Kumar
University of Virginia

How many planetary systems are there in our Galaxy? This question is being asked not only by scientists and philosophers but also by the general public. The great interest in this problem has arisen because of the obvious connection between the emergence of life and the existence of a planetary system. So far, we have no evidence for the existence of any other planetary system in the Galaxy, much less for the existence of other life forms. Yet, quite a number of scientists have stated unequivocally that our Galaxy most probably possesses many billions of planetary systems! We have not even solved the problem of the formation of our own Solar System; therefore, these statements are based neither on observational evidence nor on any satisfactory theory. The object of this essay is to discuss critically what we understand about the formation of planetary systems, and to see if such systems can be a universal phenomenon.

The Solar System

Let us begin with the Solar System, the only such system known to exist in the Universe. It exhibits the following interesting properties:

1. It is a congregation of planets moving around a single central star.

2. All the planets move around the central star in the direction of its rotation, in nearly circular and coplanar orbits. In particular, the orbit of the most massive planet (Jupiter) has a small eccentricity, $e = 0.048$ (this measures the deviation from circularity).

3. The distance of the nth planet from the central star is approximately given by the Titius-Bode relation:

A preliminary version of this essay was prepared while I was a Visiting Fellow at the Australian National University (1970–71). I am grateful to Dr. Olin Eggen for his hospitality at the Mt. Stromlo and Siding Spring Observatories.

$$r_n = (0.114)(1.89)^n \text{AU}, \tag{1}$$

where 1 AU = one astronomical unit = 93 million miles.

4. The total mass of all the planets is much smaller than the mass of the central star. In particular, Jupiter's mass is approximately 0.1 percent of the Sun's mass.

5. The total *angular momentum** of the planets is much greater (about 100 times) than that of the central star.

6. Some planets have satellite systems (moons) that orbit them in a manner similar to the way the planets move around the central star.

7. The planets may be divided into two groups: the inner terrestrial planets (Mercury, Venus, Earth, and Mars) with small masses and high densities, and the outer Jovian planets (Jupiter, Saturn, Uranus, and Neptune) with large masses and low densities. (The outermost planet, Pluto, is most probably an escaped satellite of Neptune.)

In addition to these major characteristics, let us mention that between the orbits of Mars and Jupiter move many thousands of rocky fragments, called *asteroids*. Their average distance from the Sun corresponds to $n = 5$ in equation (1), and their total mass is much less than that of our Moon. Their origin is still rather uncertain.

What Is a Planetary System?

Since there exists no clear definition of the term *planetary system* in the scientific literature, I propose the following tentative definition:

A planetary system is an assembly of planets moving in nearly coplanar orbits around *one* star, with the most massive planet having very little mass compared to the central star $(M_{mmp}/M_{cs} \lesssim 10^{-2})$ and with the most massive planet's orbit having a small eccentricity $(e \lesssim 0.1)$.

This definition ensures the dynamical stability of the planetary system over long periods of time (certainly a necessity, if life is to arise and evolve) and distinguishes the planetary system from a multiple-star system.

*Angular momentum measures the rotational impetus of a system, for it is essentially (mass) × (speed) × (distance from the center of rotation or revolution).

Note that the definition does not use many of the properties mentioned above; let us discuss those properties that we have not used here:

Property 3. No convincing reason has been given for the existence of the Titius-Bode relation, equation (1). Indeed, we would expect that the planetary distances depend upon many factors, such as the planetary masses (relative to the mass of the central star) and their orbital eccentricities; in these respects, the most massive planet is probably of predominant importance. It is doubtful that the Titius-Bode relation would apply to any other planetary system.

Property 5. In 1967 R. P. Kraft observed that all stars appear to lose angular momentum continuously from the moment of their birth (due to stellar winds and other forms of mass loss). His study revealed that young stars spin much faster than old stars and that this loss of angular momentum correlates with the age of the star. Hence, the slow rotation of our Sun (approximately one rotation per month) is probably due to its great age, and not to the fact that it has a planetary system. Consequently, theories of the origin of planetary systems would do well to attach very little importance to the present slow rotation of our Sun.

Property 6. The significance of satellite systems is not clear, since they are certainly not necessary for the formation of the planets themselves (witness Mercury and Venus). What is clear is that once we understand how planets are formed, we will probably also (as a by-product) understand their satellite systems.

Property 7. The division of the planets into two groups may be important, but at present there is no indication that such a division must obtain for every planetary system.

Let us now consider, in detail, the two criteria that were our basis for defining a planetary system: (i) the long-term dynamical stability, and (ii) the fact that we did not want to deal with a multiple-star system.

The Stability of Planetary Orbits

The general three-dimensional problem of N bodies, orbiting one another under the influence of their mutual Newtonian gravitational attraction, has never been solved. Some numerical simulations of our Solar System have been run on electronic computers; these studies show that the present planetary orbits are reasonably stable over periods of about a billion years. In particular, the Jovian planets seem to have settled into almost-commensurate orbits to ensure their stability; that is, the ratio of any two orbital periods is almost a rational fraction. For example, $(P_{Jupiter}/P_{Saturn}) = 2/5$ to within 0.5 percent and $(P_{Saturn}/P_{Uranus}) = 1/3$ to within 5 percent.

Since we are primarily interested in other planetary systems, we must somehow "solve" the general N-body problem. In analogy with the Solar System, it seems reasonable to assume that the central star and the most massive planet (Sun and Jupiter, here) will be most influential in determining the stability of the entire system. It turns out that the so-called *elliptic restricted problem of three bodies* can be handled with comparative ease. This is the two-dimensional situation where three bodies orbit in a single plane, with the body of infinitesimal mass, M_3, performing periodic (*i.e.*, closed) orbits in the presence of two very massive bodies, M_1 and M_2, in mutual elliptical orbit. Recently, P. J. Shelus and I (1970; see also P. J. Shelus 1971) have investigated the *linear stability* (*i.e.*, short-term stability) of the M_3 orbit, when both the mass ratio (M_2/M_1) and the orbital eccentricity (e) for the massive bodies is varied. We let (M_2/M_1) range from 0.001 to 1.0, where the former case corresponds to the Jupiter-Sun situation; and also let e range between 0 and 1 (its maximum value for an elliptical orbit).

The results of this study are very illuminating. The small body's orbit is stable only if it is located either very near to M_1 or M_2 or very far from both massive bodies. In particular, consider the case of the Earth, which certainly has a stable orbit in the Solar System, where $(M_2/M_1) = 0.001$ and $e = 0.048$ (Jupiter). As we (hypothetically) increase the orbital eccentricity of M_2 (Jupiter) about M_1 (the Sun), the Earth's orbit becomes unstable at $e \geq 0.5$: The Earth would either spiral into one of the massive bodies or be ejected completely from the system! Our general conclusion is this: For a given mass ratio (M_2/M_1), there exists a limiting eccentricity above which the orbit of M_3 at any position in the plane becomes unstable.

We can now understand why the mass ratio (M_{mmp}/M_{cs}) and the orbital eccentricity of the most massive planet had to be restricted in our earlier definition of a planetary system. The uncomfortable alternative—from the viewpoint of the existence of life—is to have (a) only one very massive planet orbiting the central star, (b) small planets (moons?) extremely close to or far from the two massive bodies, or (c) a multiple-star system.

The Nature of the Dark Companions

A massive planet and its central star will orbit about their mutual center of mass. It is easy to see that two bodies of equal mass must orbit a point exactly halfway between them, and that this center-of-mass point moves closer to the more massive body as we increase its mass. In the Sun-Jupiter case, where $(M_{Sun}/M_{Jupiter}) \simeq 1000$, the center of mass is located essentially at the Sun's surface. If we removed ourselves to a great distance from the Solar System, so that Jupiter could no longer be seen, we would notice that the Sun seems to "wiggle" in space with a period of about 12 years (Jupiter's orbital period); this is just the Sun's motion about the center of mass. At the distance of the nearest star, about 4.3 light-years, this wiggle would amount to only about 0.004 arc-second on the sky—totally undetectable.

Now, astrometric studies have revealed such stellar wiggling on the sky, so we know that at least six other stars are accompanied by "dark" companions (see Table 2.1). For Barnard's star, the reality

TABLE 2.1.   Data on the "dark" companions

System	Mass ratio	Eccentricity
BD + 20°2465	0.08	0.6
Lalande 21185	0.03	0.3
61 Cygni	0.04	0.5
Ci 2354	0.09	0.9
BD + 6°398	0.07	0.6
η Cas	0.04	?

NOTE: The first column gives the name of the star around which a dark companion has been detected astrometrically. The second column gives the ratio of the mass of the dark companion to that of the visible star. The third column gives the eccentricity of the orbit of the dark companion.

of the perturbation is extremely uncertain, so this star is not included in the table. Many astronomers have jumped to the unwarranted conclusion that these dark companions are planets; as I (S. S. Kumar 1964, 1969) have ofttimes said in the past (and will reiterate here), these objects are nothing but stars of very low mass that have become completely degenerate objects without going through the normal stellar evolution. Since I cannot stress this point too strongly, let me explain why it is the case.

There exists a lower limit to the mass of a main-sequence star (S. S. Kumar 1963), but this limit should not be confused with the minimum mass that a star can have at the time of its formation. The minimum mass on the main sequence, for Population I composition (about 62 percent hydrogen, 35 percent helium, and 3 percent heavier elements), is approximately 0.07 solar mass and a star with mass less than 0.07 solar mass will never go through the hydrogen-burning phase of stellar evolution. However, such a star will go through the initial collapse phase as it forms, and later it will be a fairly luminous object for periods of tens to hundreds of millions of years during the slow Helmholtz-Kelvin contraction phase of evolution. The contraction stops when the object becomes completely degenerate (see Essay 1), and then it begins to cool slowly toward the black dwarf stage. The dark companions listed in Table 2.1 are such objects. They are still far from the zero-luminosity stage, but we can not see them because they are fainter (by 5 or 6 magnitudes) than the visible primaries about which they are orbiting.

Star and Planet Formation

We have seen that there is no evidence for any planetary system other than our own. Therefore, it is necessary to embark upon theoretical speculation and logic.

What is the minimum mass that a star can have at the time of its formation? Let us attempt a preliminary answer to this question. For a just-formed star of mass M, the total energy should be negative for the star to collapse under its own gravitation. Assuming that the star is composed entirely of hydrogen, this condition gives the following relation for the minimum mass ($M_\odot$ is the Sun's mass):

$$(M/M_\odot) > 2 \times 10^{-11} T^{3/2} \rho^{-1/2}, \qquad (2)$$

where T and ρ are the temperature and density, respectively, of the star. The minimum mass of the star will be lower, the lower the temperature and the higher the density of the star. If we take $T = 3°K$ (definitely the lowest limit) and, further, if we take for ρ a value 10^5 times the mean density of the interstellar clouds, then, from equation (2), we have $M > 0.1 M_\odot$. It is likely that star formation in some regions may lead to objects with densities higher than the 10^{-18} g cm^{-3} used above, and consequently the value of the minimum mass will be lowered. For example, if $\rho = 10^{-16}$ g cm^{-3}, then $M > 0.01\,M_\odot$. We cannot push the value of ρ to much higher values, for it will be impossible to get an extended, gaseous object with $\rho \gg 10^{-16}$ g cm^{-3} out of the interstellar clouds. Therefore, we may tentatively say that the value of the minimum mass satisfies $M_{min} \sim 0.01\,M_\odot$. There is no likelihood of forming extended, gaseous objects of mass much less than $0.01\,M_\odot$ from the interstellar clouds, and therefore, one should not talk of forming planets from the interstellar medium. (Recall that $M_{Jupiter} \simeq 0.001\,M_\odot$.)

The processes of star formation are radically different from those of planet formation. A star is formed as an extended, gaseous object by instabilities in the interstellar clouds, while the planets in our Solar System have most probably been formed by *collision-accretion processes*. According to the accretion mechanism (B. J. Levin and V. S. Safronov 1960; S. S. Kumar 1967; H. Alfvén and C. Arrhenius 1970), the planets were formed by the accumulation of surrounding materials by solid nuclei or embryos formed earlier from dispersed matter around the Sun. We shall return to the comparative study of the processes of star formation and planet formation later. Since the dark companions have masses in the range $0.01\,M_\odot$ to $0.07\,M_\odot$, they cannot be anything but stars of very low mass that are slowly approaching the black dwarf stage. Incidentally, the existence of at least six low-mass dark companions within 15 light-years of the Sun strongly points to a very high space density for such objects. It seems quite certain that most of the apparently single stars are accompanied by black dwarf companions. Invisible stars of very low mass must contribute significantly to the total mass in the solar neighborhood, if the *mass function** has the form $F(M) \propto M^{-2.0}$, and if the minimum mass of a star is approximately $0.01\,M_\odot$. With $M_{min} \simeq 0.01\,M_\odot$ we can explain all the "missing mass" in the solar neighborhood in terms of degenerate stars of very low mass. If M_{min} were much lower than $0.01\,M_\odot$, there would be an "excess" mass that is not detected dynamically. This is another argument in support of our proposal that the minimum mass of a star cannot be much lower than 0.01 M$_\odot$.

---

*In the narrow mass range between M and the slightly larger mass $(M + dM)$, the number of stars per unit volume of space is $F(M)dM$; here, $F(M)$ is termed the *mass function*.

Other Planetary Systems

As we must continually emphasize, there is, at the present time, no observational evidence concerning the existence of other planetary systems in the Galaxy. We have yet to detect a second planetary system. From the theoretical viewpoint, we do not have a satisfactory theory for the formation of our own planetary system, and so we cannot realistically talk about the formation of other planetary systems and their frequency of occurrence in the Galaxy. Yet, the scientific literature is full of statements that the Galaxy possesses billions or tens of billions of planetary systems. We shall now discuss two of these proposals and the reasoning behind them.

First, let us look at the suggestions of O. Struve (1950) and of S. S. Huang (1967), who say that practically all late-type dwarfs (cool main-sequence stars) possess planetary systems. It has been known for some time that G, K, and M dwarfs have much lower rotational speeds than O, B, and A dwarfs. Furthermore, our Sun, which is a G2 dwarf, has a rotational speed of only 2 kilometers per second. Since our planetary system possesses 98 percent of the total angular momentum in the Solar System, Struve and Huang propose that other G, K, and M dwarfs also possess planetary systems. The basic assumption in this proposal is that there exists a causal relationship between the loss of angular momentum from a star and the formation of a planetary system around it. However, recent observations of the rotational speeds of G dwarfs of different ages show that these stars, and presumably K and M dwarfs also, lose angular momentum continuously via a *stellar wind*.* Our Sun has been losing angular momentum since it was born, and it is still doing so. There exists no convincing theoretical mechanism that demands that the loss of angular momentum in the early history of a star should lead to the formation of a planetary system. The angular momentum lost from a star may or may not reside in the vicinity of the star. Even if it does, it does not necessarily have to lead to the formation of a planetary system. There may be dispersed material moving around a star that may never condense into planets. Some G, K, and M dwarfs have low rotational speeds because of their old age, and it appears that the existence of low rotational speeds has nothing to

---

*The spherical outflow of gas from the very hot corona of our Sun is called the *solar wind;* this flow carries mass and angular momentum away from the spinning Sun. Stellar winds are analogous to this observed solar wind.

do with the existence of planetary systems. I pointed this out (S. S. Kumar 1967) before Kraft's observations on rotational speeds.

Other workers (*e.g.*, G. P. Kuiper 1951) have argued that the process of the formation of a Solar System is similar to that of the formation of double stars, and since double stars have a high frequency of occurrence, so also must planetary systems. This argument is most probably wrong, for the process of double-star formation is different from that of the formation of a planetary system. We have already mentioned some differences between the processes of planet and star formation. Further, the orbital characteristics (mass ratios, orbital eccentricities, and so on) of double stars are quite different from those of our Solar System. It has been estimated by C. Worley (1969) that at least 60 percent of all stars in the solar neighborhood occur in double systems. The remaining so-called single stars are most probably accompanied by faint companions (high- and low-mass degenerate stars or K and M main-sequence stars). There may also exist some genuinely single stars which are accompanied neither by other star(s) nor by planets. It is clear that the existence of a planetary system is incompatible with the existence of a double star. In general, there may be either a double-star system or a planetary system, but not both. Since practically all stars may occur in double (and multiple) systems, the occurrence of planetary systems in the Galaxy cannot be a universal phenomenon! Therefore, we may conclude that the total number of planetary systems in the Galaxy is much less than the numbers generally quoted in the literature. What the actual number is, is hard to say at this stage, but it may lie somewhere in the range 1 to 10^6. Our Solar System is by no means unique, but the total number of such systems is unlikely to be more than 10^6, even though the number of stars in the Galaxy is 10^{12} or more.

The Question of Life

Life is discussed in Essay 5—its necessities and probable forms. Life can probably arise only on the surface of a reasonably warm (but not too hot) planet in nearly circular orbit about a stable star. In addition to bearing liquid water, the surface of the planet must remain fairly uniform for (approximately) billions of years, so that there is time enough for the chance emergence and ongoing evolution of life. Our estimate here of less than a million planetary systems in the Galaxy means that the chance of other lifeforms existing in our Galaxy is low, but probably not zero. It is quite likely that the nearest planetary system with life—of any form—is more than 500 light-years distant. The probability of intelligent life in our Galaxy, other than our own, must unfortunately be considered negligible!

3. The Crab Nebula and Pulsar

Virginia Trimble

*Institute of Theoretical Astronomy,
Cambridge, England*

"On a chi-chhou day in the fifth month of the first year of the Chih-Ho Reign-Period a 'guest-star' appeared at the south-east of Thien-Kuan, measuring several inches. After more than a year, it faded away." So said the Chinese historian Toktaga in his Sung Shih, or History of the Sung Dynasty.

The people who have studied such things tell us that the day chi-chhou of the fifth month of the first year of the period Chih-Ho was July 4, 1054 A.D., and that Thien-Kuan is near the star Zeta Tauri. It is not quite clear how the ancient Chinese measured inches in the sky (although an inch, held at arm's length, is about 2°), but if we look today—or preferably tonight—a little more than a degree away from Zeta Tauri, we see a rather unspectacular faint (about 9th magnitude) nebulosity called the *Crab nebula*. Figure 7 shows it photographed with several different colored filters.

It may seem a bit surprising that, out of many thousands of nebulae in the sky and almost 600 temporary stars reported by the Chinese, two can be associated with any degree of confidence. But several factors help. First, the series of Chinese, Japanese, and Korean observations tell us that the object could be seen in the same place for more than 600 days; hence, it was neither a comet nor a meteor. Nor could it have been a variable star or ordinary nova, because there is no suitable remnant star in that part of the sky now. Thus, the 1054 event must have been a *supernova*, a rare (only about one per century occurs in our entire Galaxy) and violent kind of stellar explosion, which can blow off more than half the material of the star involved, or even disrupt it completely. One of the Chinese reports mentions that the guest star was "visible by day, like Venus," for 23 days. Since we know roughly how bright a supernova really is, this allows us to estimate the distance to the 1054 A.D. event as 4000 to 8000 light-years.

Characteristics

Modern observations of the Crab nebula, on the other hand, clearly show that it is one of a class of gaseous nebulae called *supernova remnants*. Its spectrum of emission lines superimposed on a continuum of *synchrotron radiation* (the kind of radiation emitted by high-speed electrons moving in a magnetic field), combined with the absence of any conspicuous central star, does not admit of any other interpretation. Furthermore, comparison of two pictures of the Crab nebula line emission features (like the one shown in Fig. 8), taken at different times, shows that the nebula is expanding. We can use that measured expansion in two ways: first, when combined with radial velocities measured from the spectrum, it tells us that the object is about 6000 light-years away. And second, if the expansion is extrapolated back in time, it turns out that the nebula would have been a point in about the year 1140 A.D. When we allow for the expansion having been accelerated by the pressure of high-speed electrons and magnetic fields (precisely the ones required to explain the synchrotron continuous spectrum), the most probable date for the explosion is perhaps 100 years earlier.

Thus the Crab nebula can be shown to be the remnant of a supernova which occurred in the same part of the sky, at about the same distance from us, and at about the same time as the event seen by the Chinese.

The connection between the 1054 event and the Crab nebula was first suggested in the early 1920's and confirmed by about 1940. Since that time, three or four other supernova remnants have been associated with observations of events that produced them in historic times. But no other object has been so well studied or provided so many surprises as the Crab nebula, and it remains unique in the wide variety of phenomena exhibited. It has features that move, perhaps periodically, at an appreciable fraction of the speed of light; it is an X-ray source as well as a radio source; the polarization of its optical and radio emission shows that they are produced primarily by the synchrotron mechanism—hence, the nebula must contain both a magnetic field and relativistic electrons. There may also be relativistic protons and heavier nuclei present that, if we detected them near the Earth, we would call *cosmic rays*. It was, in fact, the first radio source outside the Solar System to be identified with a particular object; the first non-Solar-System X-ray souce to be so identified; and the first object for which synchrotron radiation (now thought to be the emission mechanism for a wide variety of objects, inside our Galaxy and outside it, including quasars) was found to be important.

The nebula also contains a *pulsar*—the only pulsar to have been found at optical and X-ray, as well as radio, wavelengths. Changes impressed on the radiation from the Crab nebula pulsar (NP 0532) by the material through which this radiation passes on its way to us even tell us something about the density of the interstellar medium in our Galaxy. NP 0532 was the first pulsar to be observed, although the observers did not know what they were seeing. The radio emission from the pulsar was first recorded in Cambridge in 1964, as a small low-frequency radio source located near the middle of the larger radio source corresponding to the nebula. The optical emission from the pulsar looks just like an ordinary 16th-magnitude star, which must have been seen by many people even before the turn of the century. It is only when one makes use of electronic light-sensing devices, with good time resolution, that it becomes obvious that this star is, in fact, emitting all its light in short bursts, separated by about 0.033 second. (This period is gradually increasing, with a time scale for doubling of about a thousand years—yet another link with the 1054 event.) If the period of NP 0532 were even a factor of 2 longer, the light variation could be seen directly. Perhaps it is just as well that it cannot be; fifty years ago an astronomer who came home from an observing run with the report that he had seen a faint star which flickered regularly might well have been suspected of drinking something stronger than water with his midnight lunch.

It is a good omen for astronomy that this mélange of observations of different kinds can be put together into a reasonably consistent picture which is in good accord with our theoretical understanding of supernova explosions and their remnants. Even more propitious is the fact that much of the scenario was proposed more than thirty years ago, by F. Zwicky. He suggested in 1939 that the vast energy of a supernova explosion was derived from the formation of a neutron star and, therefore, that there ought to be a neutron star in the center of the Crab nebula. The very short pulsation period of NP 0532 implies that it is almost certainly a rotating neutron star, a striking confirmation of Zwicky's prediction.

The Crab Scenario

The story starts with an ordinary star of five or ten times the mass of our Sun, "burning" nuclear fuels as discussed in Essay 1. Initially, the star burns hydrogen to form helium. Eventually, all the hydrogen in the center of the star will be used up, leaving an inert helium core, surrounded by a thin shell in which hydrogen is still being burned. The inert core must contract and heat up, living on its gravitational energy, in order to balance the radiation from its surface. In due course, the center of the star becomes hot enough to enable the helium to burn and form carbon and oxygen, until it, too, is exhausted at the center. Meanwhile, the hydrogen-burning shell continues to eat its way outward through the star. The inert core of carbon and oxygen, in turn, contracts and heats up until further burning occurs. Thus a highly evolved star will consist of a number of layers, some inert and some burning a variety of nuclear fuels; the central core is very hot and dense, indeed, for a cubic inch of its material would weigh several tons and would contain enough heat energy to boil almost a billion gallons of water.

But this process cannot continue forever. Eventually, either one of the fuels ignites explosively (this probably blows the star apart completely), or the central core is processed by successive nuclear reactions all the way to iron, and no further reactions can extract energy from it. The core is then free to collapse rapidly, liberating large amounts of gravitational energy. While the star was evolving and growing hotter and denser at the center, its outer layers were expanding and cooling. Thus the ultimate explosion effectively occurs inside a large, cool gas cloud. Radiation trapped inside such a cloud is unstable, in the sense that instead of pushing the whole cloud out uniformly, it will break through in channels, forcing the gas into filaments between the channels. So, at this stage, we have (1) radiation that has leaked out and will, in due course, be observed as the supernova event, (2) a few solar masses of gas broken up into filamentary structures and flying off with velocities of 1000 to 10,000 kilometers per second, and (3) a one or two solar mass core which has collapsed to such a high density that the electrons and protons in it are squeezed together to form neutrons, and which is, therefore, called a *neutron star*. Such a star, once formed, is stable, and cannot contract further even when it cools off.

Most ordinary main-sequence (hydrogen-burning) stars rotate, with periods of a few hours to a few months. Many of them also have magnetic fields, with strengths ranging from one to 1000 or even 10,000 gauss. Angular momentum and magnetic flux are probably approximately conserved when a star collapses. Thus the neutron star may be spinning with a rotation period much less than one second and may have a surface magnetic field as large as 10^{12} or 10^{13} gauss when it is formed.

At this point, the scenario becomes both more complicated and less certain. It is clear that a number of different processes must occur, but there is by no means universal agreement among people working in the field on which processes are most important at which stages. Among the competing mechanisms are gravitational radiation, acceleration of cosmic rays, acceleration of high-energy charged particles in a magnetosphere around the neutron star, and electromagnetic radiation at the star's rotation frequency. All of these will drain rotational energy from the star, increasing its rotation period.

The formation of a neutron star, and its early rapid-rotation phase, are almost certainly accompanied by the radiation of gravitational waves. Although this radiation may occur as a short burst, it is probably not of the right wavelength to account for the gravitational wave events reported by J. Weber of the University of Maryland. The intensity of a neutron star's gravitational radiation will decline rapidly as the star rotates more slowly and becomes almost spherical. Even for NP 0532, the youngest pulsar known, gravitational radiation is probably negligible.

Cosmic rays, as observed at the Earth, are very-high-energy charged particles (mostly protons, but also some electrons and nuclei of elements heavier than hydrogen). It is believed that some of these particles have been accelerated to high kinetic energy in supernova events in our own Galaxy. But this process must be confined to the early phases of the explosion. Although there are undoubtedly high-energy charged particles in the Crab nebula at the present time, these will not be released into the interstellar medium until after they have lost most of their energy.

A decade or two after the supernova event, a magnetized neutron star remains, rotating 50 or 100 times a second and surrounded by an expanding cloud of filaments (already something like ten times the size of the Solar System). These filaments are mixed with high-energy charged particles and threaded by a magnetic field, partially derived from the interstellar medium swept up as the cloud expands. As the cloud expands, its density drops, and it becomes transparent to most kinds of electromagnetic radiation, from radio on up to X-ray frequencies.

The magnetic field of the neutron star is probably roughly dipolar, like the field of the Earth, an ordinary magnetic star, or a laboratory bar magnet. If the axis of the magnetic field is parallel to the star's rotation axis, then electric fields are produced, whose effect is to lift charged particles off the surface of the star into a surrounding magnetosphere and to accelerate them to high energies. If, on the other hand, the two axes are perpendicular, then the star will radiate very intense electromagnetic radiation at its rotation frequency. This frequency is so low that the radiation cannot propagate through ionized gas; so the ions and electrons are swept along by the electromagnetic radiation and again accelerated to high energies, while a cavity around the neutron star is swept clear of gas. The rate at which energy is lost from the star is the same for the two processes, and, of course, if the magnetic and rotation axes are inclined at some angle other than 0° or 90°, some combination of the two processes will occur. Thus energy is lost from the rotating neutron star, in the form of high-energy particles and low-frequency electromagnetic radiation.

As this energy is lost, the rotation gradually slows down. Sudden, discontinuous changes in the rotation period may also occur. Such changes in period (called *glitches*) have been observed for both NP 0532 and for the pulsar associated with the supernova remnant in the constellation Vela. The glitches may be due to a settling of the neutron star into a more spherical shape (a sort of starquake), or to the sudden escape of a large volume of high-energy particles from the pulsar's magnetosphere, or just possibly, to a planet sweeping close in on an eccentric orbit around the neutron star.

As a result of this highly efficient transformation of rotational energy, a continuous supply of relativistic electrons diffuses through the surrounding gas cloud or nebula. These electrons find themselves in the magnetic field of the nebula. In addition, any low-frequency electromagnetic radiation that is not used up in accelerating particles close to the pulsar will also spread through the nebula, and to a given particle at a given time, this long-wavelength radiation field will look rather like a static magnetic field. Now an electron shot into a magnetic field goes into an orbit which spirals around the magnetic field lines, and it emits electromagnetic (synchrotron) radiation as it goes. This radiation is polarized and has a frequency which depends both on the strength of the magnetic field and on the energy of the electron. Thus electrons with a wide distribution of energies, such as acceleration near the pulsar is likely to produce, will radiate at X-ray, optical, and radio wavelengths.

The realization that a pulsar can provide a continuing supply of high-energy electrons has solved one of the long-standing problems connected with the Crab nebula. The highest-energy electrons, which radiate X-rays, lose all of their energy in only a year or two. It was, therefore, puzzling that the Crab nebula, which is 1000 years old, is still an X-ray source.

With so many different kinds of things—particles, fields, and radiation—present in a supernova remnant, a wide variety of complicated interactions are bound to take place, some of which we understand less incompletely than others. For instance, the emission-line spectrum of the Crab nebula filaments is well explained by the assumption that the gas is kept ionized by the ultraviolet synchrotron radiation produced in the nebula itself. We have already mentioned that the expansion of the nebula appears to be accelerated by the pressure of the magnetic field and relativistic electrons that produce the synchrotron radiation. Among the more puzzling features are the *wisps*, regions of extrabright optical synchrotron emission near the center of the nebula, which move around with speeds up to 10 percent of the speed of light, perhaps oscillating with a period of a year or so.

One of these wisps seems to have become suddenly brighter shortly after the large discontinuity in the period of NP 0532 that occurred in September 1969. At about the same time, the electron density near the center of the nebula increased, as did the radio emission of the nebula at a wavelength of about 2 centimeters. These changes all seem to indicate that a large burst of high-energy electrons was produced at the time of the glitch. These electrons may have burst out of the pulsar's magnetosphere or been blown off a planet, rather like the tail of a comet.

Finally, as a supernova remnant ages, the nebula will merge with the surrounding interstellar medium and disappear. The pulsar's rotation period will have increased to one or two seconds, and it will emit only radio pulses and perhaps weak X-ray radiation, due to interstellar matter falling onto its surface. More than 50 of these long-period isolated pulsars have been observed (compared to two still associated with nebulae). Their average age is 10^6 to 10^7 years. Thus, if our Galaxy has been producing neutron stars at the present rate over its entire lifetime (about 10^{10} years), there should be about

100 million invisible, dead pulsars in the Galaxy. These "cinders" would represent less than 0.1 percent of the mass of the Galaxy, but they may be responsible for a large fraction of its soft X-ray emission.

The reader will notice one serious omission in this discussion: no account has been given of the pulsar mechanism itself. It is clear that the precise periodicity of the radio (or for NP 0532: radio, optical, and X-ray) pulses from these objects is accounted for by their rotation. Because of the very high intensity of the pulses and the small volume from which they come, it is also clear that they must be produced by large bunches of charged particles radiating in some coherent way. And, finally, these bunches must somehow be confined to a small sector on the surface of the neutron star, so that we see their radiation in pulses as that sector sweeps past us like a searchlight beam. But, so far, neither observation nor theory takes us further than this. The gap is not so serious for our understanding of neutron stars and supernova remnants as it might at first seem. Although the existence of the neutron stars proposed so long ago by Zwicky was first demonstrated by their pulsed radio emission, much less than 1 percent of the energy lost by the rotating star goes into that pulsed emission.

Violent and Active Extragalactic Objects

Since the explanation of supernova remnants in terms of rotating magnetic neutron stars and their products has been so successful (at least by astrophysical standards), a good deal of effort has recently gone into trying to explain the behavior of quasi-stellar objects and the active centers of galaxies in some analogous way; the Crab nebula is thereby taken as a prototype for high-energy objects throughout the Universe. The case for this analogy is strengthened by the many properties that some *quasi-stellars* and *active galactic nuclei* have in common with the Crab nebula. Among these are:

1. They are synchrotron radiation sources and must, therefore, contain magnetic fields and relativistic electrons.

2. They radiate a large fraction of their energy at short wavelengths and must, therefore, have a large fraction of their relativistic electrons accelerated to very high energies. The Crab nebula emits about 90 percent of its radiative energy at X-ray frequencies.

3. They have explosive events, which (like the wisps in the Crab nebula) propagate at speeds close to the speed of light.

4. They have preferential directions, defined by jets sticking out of their centers or by the axes of radio sources, which consist of two components on opposite sides of their parent galaxies. In the Crab nebula, the wisps always move along the major axis of the elliptical shape of the nebula, and this is also the axis of polarization of the nebula.

5. Their spectra show emission lines, which must come from gas in filaments at rather similar temperatures and densities to those which prevail in the Crab nebular filaments.

6. At least one quasar (3C 273; see Fig. 16) probably contains a small low-frequency radio source, and the radio emission from NP 0532 was first detected as such a source.

7. At least one quasar (3C 345) may exhibit periodic variations in brightness, which would be analogous either to the pulses from NP 0532 or to the apparently periodic behavior of the wisps in the Crab nebula.

8. They exhibit energetic behavior very far away from what appears to be the source of the energy: the diameter of the Crab nebula is more than 10^{12} times the diameter of its central neutron star. And some radio galaxies have small radio-emitting components separated from their parent galaxies by distances up to 10^4 times the size of the galaxy.

9. Many of them vary in intensity, as does the Crab nebula at some wavelengths.

Two kinds of models are possible: the quasar or active nucleus might be assumed to contain a single, massive (*e.g.*, 10^8 solar masses), rotating, magnetized object or a large number of separate, pulsar-sized objects. The observations do not, at the moment, definitely favor one or the other of these possibilities. But the general idea of rotating magnetic objects acting as sources of low-frequency electromagnetic waves (which can, under some circumstances, travel very great distances without losing their energy) and high-energy particles may be the correct explanation of many properties of active extragalactic objects in the Universe.

Fig. 1. Earth from Gemini 4. Photograph taken at 100 miles altitude, looking west-ward from the Hadramaut Plateau across the Gulf of Aden to Somalia (Africa). The very thin band of atmosphere hugs the curved horizon in the distance. (NASA photograph)

Fig. 2. The lunar crater Eratosthenes. This impact crater, with its terraced walls and central peaks, is located at the edge of Mare Imbrium near the center of the visible side of the Moon. Note the raised rim and evidence of distantly scattered rubble. (NASA photograph; from Apollo 12)

Fig. 3. Earth seen from the Moon. This crescent Earth was photographed over the Moon's limb by the unmanned Lunar Orbiter 1. North is toward the right; the eastern coast of North America is at the top right, Antarctica is at the left, and southern Europe is at the right. (NASA photograph)

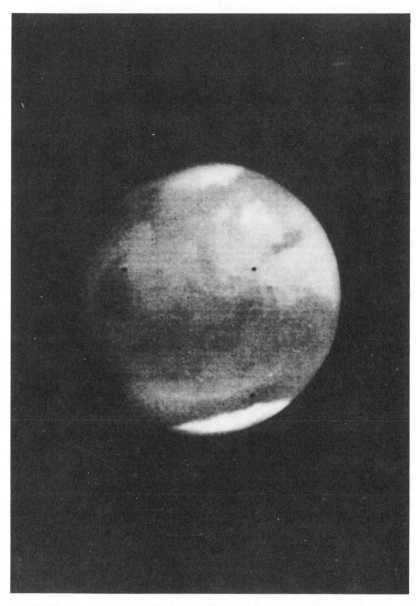

Fig. 4. Mars as photographed by Mariner 7. As the unmanned Mariner 7 spacecraft approached the planet Mars in August 1969, it took this television picture at a distance of 536,000 miles. Note the southern polar cap (*bottom*; it is spring in the Southern Hemisphere), the northern "clouds" (*top*), and the numerous surface markings. (NASA photograph)

Fig. 5. The giant planet Jupiter. Of this largest and most massive planet of the Solar System, we can see only the atmosphere of hydrogen, helium, and some molecules. Note the (colored) bands running parallel to the equator, and the famous Great Red Spot in the south (_bottom_; in telescopes the image is usually inverted—top to bottom, and left to right). (Hale Observatories photograph)

Fig. 6. The ringed planet Saturn. Next in size to Jupiter, Saturn is the only planet in the Solar System with rings. These rings, probably consisting of "snow" and small chunks of "ice" and rock, are only a few miles thick but spread from 40,000 to 85,000 miles (from the planet's center) in the equatorial plane. (Hale Observatories photograph)

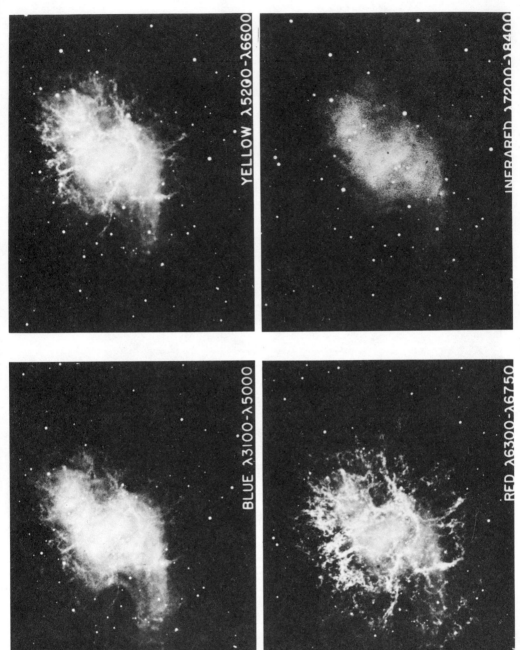

YELLOW λ5200-λ6600

INFRARED λ7200-λ8400

BLUE λ3100-λ5000

RED λ6300-λ6750

represent the Crab nebula. These rapidly expanding gaseous remains of a star which exploded in 1054 A.D. are seen in

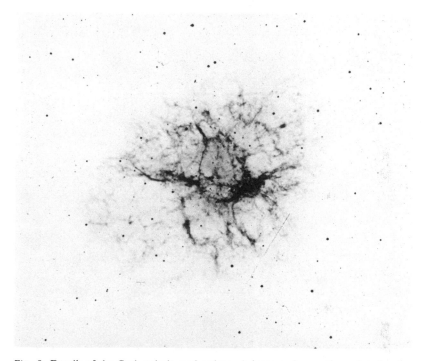

Fig. 8. Details of the Crab nebula and pulsar. A large-scale negative print showing the filaments of the Crab nebula in OII line emission. At the center are two stars; the one on the lower right is the pulsar, which excites the entire nebula while blinking on and off about 30 times each second. (Hale Observatories photograph)

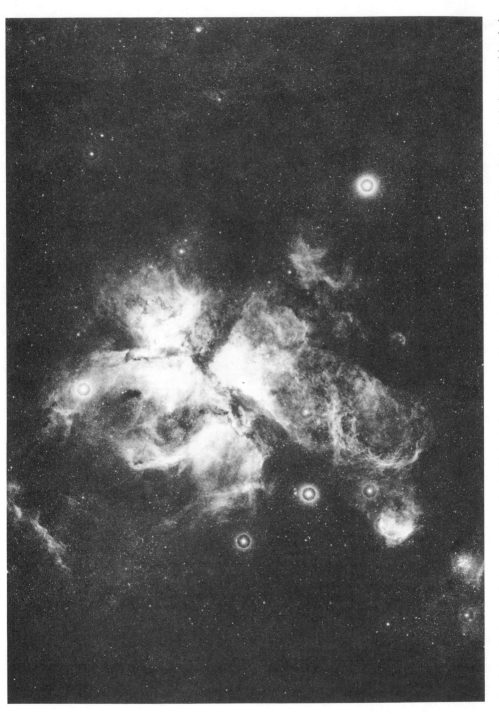

Fig. 9. Eta Carina nebula. An enormous cloud of interstellar gas, caused to glow by nearby stars, this bright emission nebula is crossed by dark lanes of dust. The constellation Carina is not visible at high northern latitudes. (Cerro-Tololo Inter-American Observatory photograph)

Fig. 10. Filamentary nebula in Cygnus. This glowing interstellar cloud (NGC 6960) seems to lie at the boundary of an obscured region (*top*) in the northern constellation Cygnus (The Swan). The nebula is expanding, and may be the remains of a star which has exploded. (Hale Observatories photograph)

4. Chemistry between the Stars

Lewis E. Snyder
University of Virginia

About two hundred years ago, the famous astronomer William Herschel found the first interstellar clouds. He described them to his sister as "holes in the sky" through which one could see to infinity. Some time later, these dark patches were recognized to be interstellar clouds (rather than holes), and after many more were discovered by optical astronomers, they were accepted as legitimate occupants of our Galaxy. But it was not until 1937 that astronomers began to understand the chemical composition of interstellar clouds. T. Dunham observed an absorption spectrum at 4300 Å— unrelated to the stars which he was observing—which turned out to be the characteristic absorption of the interstellar chemical radical CH. This discovery marked the birth of the formal study of interstellar chemistry. (Some historians might contend that interstellar chemistry began with the discovery of interstellar atomic absorption spectra, but here I will make the old distinction that the study of atoms is physics and the study of molecules is chemistry.) Soon afterward, optical astronomers detected interstellar CN and CH^+, and in the late 1930's and early 1940's they began to form the basic ideas of what constitutes the chemistry of the interstellar medium.

The Old Chemistry

A typical interstellar cloud was thought to contain a gas composed of single atoms (primarily hydrogen), of the simplest molecules having two atoms (such as H_2, CH, CH^+, and CN), and of grains consisting of methane (CH_4), ammonia (NH_3), and hydrogen (H_2) ices. It was apparent (at the time, anyway) that nothing much more complex than a diatomic gas could be expected because the molecules had to form by two-body association of their atoms and, in addition, any complex polyatomic molecule that happened to form (in blithe violation of accepted theory) would immediately be destroyed by the omnipresent ultraviolet radiation. Thus, for thirty years not much happened in the theory of interstellar chemistry, and optical astronomers were content to survey the Galaxy, collecting and cataloging their observational data on the location of interstellar clouds containing CH, CH$^+$, and CN.

This situation might never have changed except for the rapid development of radio astronomy in the past twenty years. Radio astronomers first began to study spectral lines in 1951 with the detection of the 21-centimeter line of atomic hydrogen. Twelve years later OH became the first interstellar molecule detected by radio techniques and the fourth molecule found in the interstellar medium. It was identified by its characteristic 18-cm absorption spectrum against the radio source Cassiopeia A. Thus some twenty-six years elapsed between the detection of CH, the first interstellar molecule, and OH, the fourth. No significant changes had occurred in the theoretical ideas about the chemistry of the interstellar medium during this time, but a powerful new method for interstellar spectroscopy—radio astronomy—had been born. Although its value probably was not fully appreciated by astronomers at the time, we shall see that radio spectroscopy is in many ways superior to optical spectroscopy for studying the chemical composition of the interstellar clouds.

The New Chemistry

The year 1968 began as a rather ordinary one for astronomical spectroscopists. Radio spectroscopists were busy surveying 21-cm hydrogen lines and OH lines. The newest development in radio spectroscopy was the study of recombination lines of atomic hydrogen (1965) and carbon. An infrared search for interstellar ice (frozen H_2O) by a group from Berkeley had produced no promising detections; this result was taken by some astronomers as a convincing argument (convincing from an astronomical standpoint, anyway) against interstellar polyatomic molecules. After all, if there is no detectable ice, then there is "undoubtedly" no detectable H_2O vapor; if water cannot be found, then probably nothing more complex can be present in significant quantity in interstellar clouds.

Meanwhile, against the tenets of theoretical astronomy, a group of physicists and engineers began to search for interstellar ammonia using a new 21-foot millimeter-wave telescope at the University of California in Berkeley. In late 1968 they announced the first detection of interstellar ammonia (NH_3) at 1.26 cm toward the galactic center (ref. 1). The reported measured abundance of ammonia, given in terms of *projected density*, was greater than 10^{16} molecules per square centimeter. This projected density represents the number of molecules contained in a column of gas 1 cm × 1 cm square and (theoretically) of infinite length. In practice, projected density is the quantity measured directly by the astronomical spectroscopist, since there usually is no way to measure the depth of an interstellar gas cloud. For comparison, one cubic centimeter of air at sea level has a projected gas density about 10^6 times greater than the projected density of NH_3 in the galactic center clouds. Astronomical projected densities are often converted to the more familiar measure of density, number of molecules per cubic centimeter, by indirectly determining the cloud depth and dividing it into the measured projected density. For example, if a cloud containing a projected density of 10^{16} molecules per cm^2 is indirectly determined to be 1 light-year in depth (or about 10^{18} cm), the molecular density works out to be 0.01 molecule per cm^3. Thus interstellar ammonia is not very dense by terrestrial standards, but the initial detection proved that at least one polyatomic molecule could exist in detectable quantities in the interstellar clouds—in direct contradiction to all observational evidence accumulated in astrochemistry up to 1968.

If ammonia had been the only polyatomic molecule found, nothing very disruptive might have happened to the theories of interstellar chemistry. After all, NH_3 was detected only in the galactic center, and it was quite possible that something strange was happening in the center of our Galaxy, which is a fairly mysterious place, anyway. Perhaps some of the ammonia ice had evaporated from interstellar grains at a rate faster than it was being destroyed by ultraviolet radiation. Arguments such as these were beginning to be heard among astronomers when the detection of emission signals from interstellar water vapor (H_2O) at 1.35 cm was announced by the Berkeley group (ref. 2). Water was the second polyatomic molecule found in the interstellar clouds, and it was detected initially in the galactic center and two other sources outside the center—destroying the galactic center "polyatomic uniqueness" theory soon after it was suggested. To add to the growing excitement, the intensity of the H_2O emission from the radio source W49 was at least 2000 times stronger than it was supposed to be. Any spectral emission line with an intensity which is much greater than can be produced by typical interstellar temperatures and abundances is said to be in *maser emission*.* Maser emission was not new to astronomy; it was first observed in OH as early as 1965. However, one new property of the interstellar water line—the "rain in space"—is that in some radio sources it appears to be liberating large (by spectroscopic standards) amounts of energy, in some cases on the order of 10^{25} watts. Since the 1.35-cm H_2O line represents only one of hundreds of potentially energetic but undiscovered lines in the astronomically available spectrum it is easy to see that some interstellar molecules may provide an important cooling mechanism when interstellar clouds undergo gravitational contraction. Other properties of the "rain in space" are also of interest to astrophysicists. Some of the sources have spectral features whose intensity changes during periods of about ten days or even less. At least one source, W49, has H_2O lines whose velocity differs by about ±200 kilometers per second from the velocity of the Sun in the Galaxy. Recently, astronomers have used radio interferometry to determine the sizes of several regions emitting the 1.35-cm water line and have found that their dimensions are comparable to our Solar System. Because of these unusual properties, water vapor emission is of great astrophysical interest; but unfortunately interstellar H_2O has been found only in galactic regions where the excitation conditions cause maser emission. The maser emission makes it almost impossible to determine the number of water molecules that give rise to a spectral line. Without this abundance information it is hard to make measurements relevant to processes of the interstellar chemistry; so we might say that interstellar H_2O is useful for astrophysics but not very astrochemically revealing.

The astrochemistry situation started improving in March 1969, when we first detected the interstellar formaldehyde (H_2CO) absorption line at 6.21 cm (ref. 3). We used the 140-foot radio telescope of the National Radio Astronomy Observatory in Green Bank, West Virginia. Formaldehyde was the first polyatomic organic molecule found in the interstellar clouds, and we were delighted that—unlike NH_3 and H_2O—it could be observed in many different regions of our Galaxy. For example, on the very first ob-

*The maser effect takes place when some mechanism excites many more molecules into a high-energy state than is normal for a gas in simple thermal equilibrium. These molecules can then return to a lower-energy state by emitting radiation of a characteristic wavelength; hence, we "see" superstrong radio emission lines.

serving run 60 percent of the galactic sources observed showed
detectable H_2CO absorption signals. It was readily apparent that
large regions of the Galaxy were filled with clouds containing formaldehyde at densities comparable with that of OH; thus we were
optimistic that we could learn a great deal about the chemical
physics of the interstellar clouds through abundance measurements
of H_2CO. For example, the problem of determining the projected
density (NL) of H_2CO in a given cloud appeared to be as simple as
evaluating

$$NL = 1.74 \times 10^9 \left(\frac{T_G}{T_B}\right) \int T_A \, dv \, \frac{\text{molecules}}{\text{cm}^2}, \qquad (1)$$

where the integration over the product of T_A (the antenna temperature; or intensity) and dv (the width of the H_2CO absorption line) is
simply the area enclosed by a given formaldehyde line. T_G is the
excitation temperature of the gas that characterizes the relative gas
population distribution between the two quantum levels defining the
absorption transition, and T_B is the brightness temperature of the
background source that is beaming radio energy through a particular formaldehyde cloud. By evaluating equation (1), we were able to
compute projected densities as high as 6×10^{14} formaldehyde molecules per cm^2 in the more dense clouds. However, we soon learned
that this number was probably only a very conservative lower bound
on the true projected density. In the observing session immediately
after the H_2CO detection, we decided to concentrate on dark
nebulae, the optically visible and relatively isolated interstellar
clouds of the type which Herschel had described as "holes in the
sky." We hoped that excitation conditions would be such that we
would find the 6-cm formaldehyde line in emission in those dark
clouds that were already known to have normal OH emission. Instead, curiously, we found very weak H_2CO absorption signals
(ref. 4). This was surprising because no corresponding background
continuum sources could be detected. Yet there had to be a continuum source with a temperature exceeding that of the absorption
lines detected in the dark nebulae. After some fruitless searching, we
concluded that the formaldehyde in the dark nebulae had to be absorbing radiation from the 3°K universal microwave background.
Thus the formaldehyde had to be cooler than 3°K; in fact, the excitation temperature T_G in equation (1) had to be less than 1.8°K.
The dark cloud H_2CO absorption is then just the inverse of maser

emission; such an *inverse maser* is a "super absorber," since the absorption line is much stronger than it should be for the given amount of gas present in the cloud.*

Several mechanisms have been proposed to explain the formaldehyde inverse maser action in interstellar clouds. The main "pumping" mechanisms that have been suggested include collisions with electrons or hydrogen; the effect of shock waves on H_2CO; and H_2CO superabsorption in dark nebulae caused by the energy from the microwave background radiation itself. Unfortunately, all three of these proposals have some drawbacks, but it is expected that observations of other H_2CO transitions (in addition to the 6-cm line), as well as precise measurements of the spectral energy distribution of the $3°K$ background, will solve a few of the theoretical problems connected with H_2CO pumping mechanisms. For example, it would be satisfying to find a pumping mechanism which explains fairly normal OH emission in coexistence with H_2CO superabsorption in dark nebulae—without placing stringent requirements on the density or temperature of the clouds.

Needless to say, our original projected densities (with a maximum of approximately 10^{15} molecules per cm^2) were early casualties of our detection of dark cloud H_2CO absorption and the resulting flurry of proposed pumping mechanisms. For example, collisional pumping reduces T_G in equation (1) by an order of magnitude, thereby reducing all projected densities by a like amount. On the other hand, a radiation-trapping model has just the opposite effect, and if it holds, all of our abundances should be raised by one order of magnitude. By the way, our initial values of projected density that were found from equation (1) fortuitously tend to fall about midway between the extreme values predicted from various pumping models; so they may be regarded as average abundance estimates.

It turns out that interstellar formaldehyde really gave the first solid clues that astronomers were dealing with a new chemistry. In spite of problems with abundance determination, it was obvious that H_2CO is abundant, whether the maximum projected density is 10^{14}, 10^{15}, or 10^{16} molecules per cm^2; it was also readily seen that H_2CO is widespread throughout the galactic gas clouds. From this it is not hard to conclude that the formation of interstellar organic polyatomic molecules does not require extremely unusual conditions;

*The inverse maser effect occurs when some mechanism depletes the number of molecules in an excited (high-energy) state, thereby "refrigerating" the gas cloud relative to the expected thermal equilibrium condition.

hence, interstellar chemical evolution must include more sophisticated processes than had been previously assumed. How far can the new chemistry go and what are its implications? How are interstellar molecules formed? Let us discuss these questions and others.

Quantitative Analysis of the Interstellar Clouds

Since the detection of formaldehyde, seventeen new interstellar molecules have been reported. The word *reported*, rather than *discovered* or *detected*, is used deliberately, for two reasons. It is very probable that several additional interstellar molecules have been found but not reported because the astronomical spectroscopists involved believe that more data should be obtained before formally announcing the detection. It is also highly possible that at least one (but not very many) of the molecular detections already announced will turn out to be spurious. To explain the latter situation to the non-spectroscopist, it should be mentioned that many of the molecular identifications are initially made on the basis of a single line. The interstellar line frequency is compared against either laboratory measurements or theoretical calculations after correcting for the Doppler shift caused by the velocity of the molecular cloud. The reliability of such an identification can be as high as 99 percent, given the density of known interstellar lines and present radio receiver sensitivity. However, in order for a particular molecular identification to be absolutely certain from a spectroscopic standpoint, it is necessary to detect other transitions of the molecule at different frequencies. This can be accomplished in one of three ways:

1. By detecting spectral lines from the same molecular isotope with quantum levels completely different from those of the first transition.

2. By detecting a spectral line from a less common molecular isotope.

3. By identifying the hyperfine splitting or some other pattern, if it exists, that is unique to the spectral line in question.

If a molecular identification has not passed at least one of these tests, astronomical spectroscopists understand that its identification is still open to some question.

Table 4.1 lists the twenty-four interstellar molecules reported as

TABLE 4.1.   The Twenty-Four Interstellar Molecules Reported as of January 1972

Inorganic (5)		Organic (17)			Other (2)
Diatomic	Triatomic and higher	Diatomic	Triatomic	Tetratomic and higher	
OH—hydroxyl radical	NH$_3$—ammonia	CH	HCN—hydrogen cyanide	H$_2$CO—formaldehyde	X—ogen
H$_2$—molecular hydrogen	H$_2$O—water	CH$^+$	OCS—carbonyl sulfur	H$_2$CS—thioformaldehyde	HNC—hydrogen isocyanide
SiO—silicon monoxide		CN—cyanogen radical		HC$_3$N—cyanoacetylene	
		CO—carbon monoxide		CH$_3$OH—methyl alcohol	
		CS—carbon monosulfide		HCOOH—formic acid	
				HCONH$_2$—formamide	
				HCOCH$_3$—acetaldehyde	
				HNCO—isocyanic acid	
				CH$_3$CN—methyl cyanide	
				CH$_3$C$_2$H—methylacetylene	

of January 1972.* Since there is not space to discuss each molecule individually, the main points of astrochemical interest will be mentioned.

The first group listed in Table 4.1 are the five inorganic molecules found in the interstellar clouds. All of these except molecular hydrogen were detected by radio astronomers. Even though H_2 is the most common interstellar molecule in the Galaxy, its relatively poor spectroscopic properties made it impossible to detect until June 1970, when a rocket-mounted ultraviolet spectrometer recorded eight transitions of the Lyman resonance series between 1000 Å and 1120 Å. Silicon monoxide is the first (and so far the only) interstellar silicon compound detected.

From the table it is easily seen that the organics form the largest group of interstellar molecules and, furthermore, that the polyatomic molecules dominate this group. So far, many of the large organic polyatomic molecules have only been found in the galactic center clouds, but more sensitive radio receivers will probably find them to be distributed throughout the Galaxy (in the organic group, only CH, CH^+, and CN were not discovered by radio spectroscopists, and radio transitions of CN have been observed).

Several molecules in the organic group are particularly interesting because they seem to suggest trends in the interstellar chemistry that may be important in understanding molecular formation processes. For example, we can form overlapping groups if we classify the organics according to their suggestive constituent fragments as follows:

$$\text{carbon monoxide: } CO, \ H_2CO, \ OCS, \ HNCO, \ HCOOH,$$
$$HCONH_2, \ HCOCH_3$$
$$\text{methane } (CH_4): \ CH, \ CH^+, \ CH_3OH, \ CH_3CN,$$
$$HCOCH_3, \ CH_3C_2H$$
$$\text{acetylene } (C_2H_2): \ HC_3N, \ CH_3C_2H.$$

The above crude groupings are based on molecular structure considerations and illustrate some of the apparent trends that seem to be emerging—carbon monoxide chemistry, methane chemistry, and acetylene chemistry. One could make a strong case for a CN chemistry, an OH chemistry, and an ammonia chemistry as well, but the important point to note is that most of the complex polyatomic molecules found to date are combinations of the simpler poly-

---

*For a list of the reported transitions of each interstellar molecule (as of January 1972) see reference 5.

atomic subgroups, such as a CH_3 fragment from methane or a C_2H fragment from acetylene. It is possible to extend this argument a bit more and notice that CS and SiO are both analogs of CO from the chemical bonding standpoint; but CH and CH^+ are related much more weakly. The existence (or absence) of such trends will play an important role in helping us to understand the direction of chemical evolution in our Galaxy.

The last column in the table, classified as "Other," contains two molecules which may be the most chemically interesting of all interstellar molecules found. We accidentally discovered the X-ogen line during a search for the carbon-13 isotope of HCN (ref. 6). Its rest wavelength is 0.336 cm, and although the line is molecular in origin, it cannot be identified with any known molecule; hence its name X-ogen, which stands for "of unknown extraterrestrial origin." W. Klemperer has pointed out that the molecular ion HCO^+ could be responsible for the X-ogen line, but it probably will be several years before this suggestion can be verified by laboratory measurement because of the difficulty of measuring the microwave properties of molecular ions. X-ogen is a molecule which is not rare in the interstellar medium, but it is almost unknown in terrestrial laboratories. So far, all attempts to detect the carbon-13 isotope of HCO^+ in the interstellar clouds have failed.

The second molecule, interstellar HNC (hydrogen isocyanide), indirectly owes its detection and identification to the X-ogen discovery. Following the X-ogen detection at a frequency of 89.190 GHz ($\lambda = 0.336$ cm), G. Herzberg suggested HNC as a possible identification. HNC, a rare isomer of HCN (hydrogen cyanide), has an interesting history; chemists have sporadically followed its elusive trail for years in terrestrial laboratories. In the 1930's it was recognized that H could bond to CN in either of two ways; thus it was hypothesized that all normal samples of gaseous hydrogen cyanide should contain a small percentage of hydrogen isocyanide. HNC was expected to constitute less than 5 percent of the mixture since most of the gas was expected to be in the form of the most stable isomer, HCN. Exhaustive laboratory searches for HNC failed to produce a positive detection, although some brief excitement was caused when the carbon-13 isotope of hydrogen cyanide was mistakenly identified as hydrogen isocyanide. Thus, chemists gradually lost interest in trying to identify gaseous HNC and turned to other things. Finally, in the early 1960's the ice form of HNC was successfully generated in an argon matrix at 4°K (ref. 7). In spite of this, pursuing Herzberg's suggestion was not particularly easy since no laboratory data existed for gaseous HNC. From the solid-state data and from considerations of molecular orbital theory, it appeared that the molecule had to have a linear structure. Various computations using different combinations of possible bond lengths were tried until several trends became apparent. First of all, it became clear that most of the reasonable combinations tried for the HN and NC bond lengths yielded frequency predictions slightly higher than the X-ogen frequency at 89.190 GHz. Also, most of the predictions were clustered in a relatively narrow range, roughly between 90 GHz and 91 GHz. Equipped with these frequency predictions, we then detected a molecular transition at 90.665 GHz in the interstellar gas clouds W51 and DR21. The actual bond lengths of NH and NC can be calculated by working backward from the astronomically determined rest frequency; excellent agreement is found when these calculated bonds are compared with laboratory measurements. Thus the emission signal at 90.665 GHz is possibly the first spectroscopic evidence for the existence of gaseous HNC. Of course, any such identification must be properly qualified until other transitions of the molecule are found. So, in the interest of scientific caution, HNC is listed under "Other" until the identification is observationally proved; then HNC can properly take its place under "Organic Triatomic," without the qualifying statements.

The Cloudy Future

What future developments can we expect from the study of interstellar molecules? The detections of X-ogen and HNC emphasize the unexploited role of the interstellar gas clouds as chemical generators of exotic molecular species. Molecules that have been predicted to exist by quantum mechanics, but not found in terrestrial laboratories because of their short lifetimes, may be easily detectable in the interstellar clouds. In theory, it should be possible to determine a complete set of structural parameters for nonterrestrial molecules such as HNC by using only astronomical frequency measurements. In practice, once a particular molecular transition frequency has been assigned through astronomical spectroscopy, it may become possible to generate and measure the same molecule in the laboratory. Thus spectroscopic information obtained from nonterrestrial molecules is expected to give chemists new information about the quantum mechanics of molecular bonding in terrestrially unstable species.

From a more astronomical standpoint, easily detectable molecules such as CO and H_2CO will be used to map regions of our Galaxy in much the same way that the 21-cm line of atomic hydrogen is used today. New information will be gleaned about the thermodynamic structure of the Galaxy by using interstellar molecules as local thermometers in the galactic gas clouds.

H. Urey once pointed out that the chemistry of the interstellar medium is—by volume—the common chemistry of our Galaxy, while terrestrial chemistry is exotic in comparison since terrestrial conditions probably exist nowhere else in the Galaxy. Thus, understanding the astrochemical reactions responsible for interstellar molecules is of special importance.

Let us consider a simple macroscopic model of the astrochemistry of a gas cloud. We assume that interstellar "dust" grains within the cloud provide the necessary catalytic surfaces for molecular formation. Ultraviolet radiation (*e.g.*, from nearby stars) impinging upon a grain can provide the energy needed for surface exchange reactions and subsequent molecular release from the grain surface. The grains in an interstellar cloud probably act as a radiation filter, attenuating incoming ultraviolet energy E_0 (average energy per photon) according to the simple law of exponential decay:

$$E(r) = E_0 e^{-\alpha(R-r)} \qquad (0 \le r \le R); \qquad (2)$$

R is the radius of the (spherical) cloud, r represents distance from the cloud's center, and α (units = length^{-1}) is an average value for the absorption coefficient of the cloud. Suppose that the rate of molecular release from a grain becomes significant when the ultraviolet energy E exceeds a certain threshold value E_T. We would then expect a given region of the cloud to contain molecules if $E(r) \ge E_T$ is satisfied. But we also know that molecules can be destroyed by this radiation; so let us make the reasonable assumption that photodestruction is the main mode of radiative destruction; a given molecular species is destroyed when E exceeds the photodestruction energy E_P. An interstellar cloud immersed in an ultraviolet radiation field E_0 would then be expected to contain CO, for example, if at point r the condition $E_T \le E(r) \le E_P(\text{CO})$ holds. Of course, each molecular species has its own photodestruction energy, and the value of E_P for CO is much greater than that for many other interstellar molecules. The resulting condition for the presence of any molecular species X at an interior point r in the cloud is $E_T \leq E(r) < E_P(X)$, where the value of $E(r)$ is determined by the effective values of E_0 and α in equation (2). Using laboratory data, the hierarchy of E_P values for three of the known interstellar molecules can be written as follows: $E_P(CO) > E_P(HCN) > E_P(H_2CO)$. In a given cloud of radius R containing CO, HCN, and H_2CO, it is most probable that all three molecules would exist in a central core defined by $0 \leq r < r(H_2CO)$ for which the ultraviolet radiation satisfied $E_T < E(r) < E_P(H_2CO)$. In the annular region surrounding the central core, defined by $r(H_2CO) < r < r(HCN)$, only HCN and CO would be expected since the ultraviolet energy $E(r)$ satisfies the condition $E_P(H_2CO) < E(r) < E_P(HCN)$. Finally, in the outer region of the cloud, defined by $r(HCN) < r < r(CO) \leq R$, only CO could be expected since the ultraviolet radiation photodissociates H_2CO and HCN shortly after they are released from the grain surfaces. Diagram 4.1 illustrates the resulting shell structure expected for a homogeneous, spherically symmetric cloud illuminated uniformly by an external ultraviolet radiation field.

If this model of molecular formation is valid, the ultraviolet energy density determines both the creation and destruction rates for interstellar molecules, and the interstellar grains have the dual role of providing catalytic surfaces and acting as elements of an ultraviolet energy filter. What observational programs can be used to test this theory? Obviously, spectral-line cloud maps could be used to compare the spatial extent of the various interstellar molecules. But it is not enough to observe that, for example, CO extends farther than HCN in the Orion nebula, because the interstellar excitation conditions required for spectral detection usually are not the same for different molecular species. In general, the response of a given molecular species to collisions and radiation is a function of both the dipole-moment matrix element,* $|\mu|$, and the resonant frequency of absorption or emission, ν. Thus, if two molecular species, say A and B, can be found such that the approximate equality

$$\frac{|\mu_A|^2}{|\mu_B|^2} \approx \frac{\nu_B}{\nu_A} \qquad (3)$$

is satisfied, then in an optically thin interstellar cloud with negligible

*This quantity is calculated from quantum mechanics, and it determines the probable rate at which radiation is emitted from the molecule at the specified frequency. Here $|\mu|$ denotes the rate from the lower to the upper energy level.

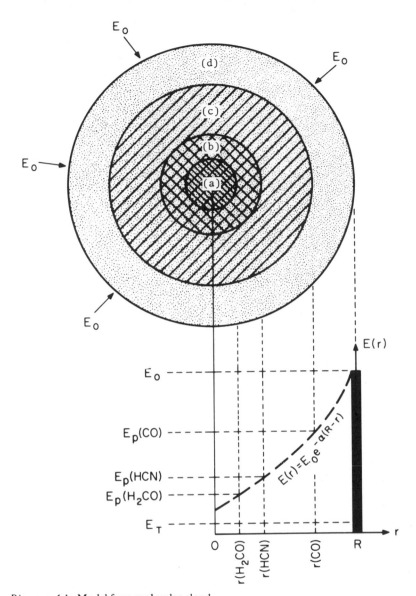

Diagram 4.1. Model for a molecular cloud

(a) In the core, $E_T < E(r) < E_P(H_2CO)$; H_2CO, HCN, and CO are released from the grain surfaces and are not photodissociated by $E(r)$.

(b) In the next shell, $E_P(H_2CO) < E(r) < E_P(HCN)$; H_2CO is photodissociated in this region, but HCN and CO exist.

(c) In the outer shell, $E_P(HCN) < E(r) < E_P(CO)$; only CO can survive the ultra-violet radiation in this region.

(d) None of the molecules we are considering (H_2CO, HCN, CO) can exist here; the "dust" grains filter the incoming ultraviolet energy from E_o down to $E_P(CO)$.

radiation density near the resonant frequencies ν_A and ν_B we would expect that a point-by-point map comparing the spectral line intensities from the two molecules could show whether spatial extent is a function of molecular photodissociation energy.

Finally, the future of astrochemistry hinges upon our understanding the direction of chemical evolution in the Galaxy. At present, we think that the interstellar chemistry prefers reactions producing organic molecules. This may be only an observational bias in astronomical search techniques, since of the million or so terrestrial molecules known to man, about 500,000 are organic compounds. But it is strange that the observed abundance ratios of several interstellar organic molecules appear to be higher than the Solar System ratios. In addition, it seems that the interstellar reactions producing the organic molecules H_2CO, CO, HCN, and CN are more efficient than those producing the inorganic molecules OH and NH_3. It is possible that amino acids (such as Glycine and Alanine) and even more complex molecules are generated by astrochemical reactions in interstellar clouds. Such a mixture has been called "life soup"; hence, the associated interstellar cloud could be called a "biocloud." Today the arguments against the existence of bioclouds are not as convincing as they once were; however, whether they exist or not, it is evident that an understanding of the direction of galactic chemical evolution is crucial if we are to determine the role of interstellar molecules in the origin of primordial life (see Essay 5).

Epilogue

Following the discovery of interstellar OH, five years passed before the next interstellar molecule, NH_3, was detected in 1968. In 1969, H_2O and H_2CO were found; in 1970, eight new molecules were discovered. By mid-1971 (at the time of this writing) nine additional molecules had been reported, with more in sight. The molecular detection rate is increasing so rapidly that some astronomers and astrophysicists (primarily those who have not found any new molecules) are privately proposing a slower rate of new detections, until the present data can be arranged in an orderly fashion. Let us examine some of the background that has led to the present state of disorder and the resulting feeling of despair among some astronomers as their orderly picture of the galactic chemistry is destroyed.

We have seen that from the late 1930's until about 1968, the accepted dogma among many astronomers seems to have included the idea that only diatomic molecules could survive in the interstellar clouds. This idea probably (but not logically) was a result of the failure of optical spectroscopists to detect any interstellar molecules beyond CH, CH$^+$, and CN, as well as the failure of radio spectroscopists to find radio lines of CH, H$_2$$^+$, and a few other expected diatomics. It turns out that the optical observers have valid reasons for not being able to detect complex interstellar molecules.

The typical molecular excitation energy required for an optical absorption line may be written as

$$E = E_{el} + E_{vib} + E_{rot}, \qquad (4)$$

where E_{el} is the energy for electronic excitation and E_{vib} and E_{rot} are the much smaller energies for vibrational and rotational excitation of the molecule. E_{el} may be on the order of 13,000 cm^{-1} (8000 Å) or greater, while E_{vib} is typically 1000 cm^{-1} to 5000 cm^{-1} above the zero-point vibrational level and E_{rot} is generally much less than 500 cm^{-1} for most of the interstellar molecules detected to date. Therefore, to observe optical transitions of interstellar molecules we need a source furnishing an "energy" of 13,000 cm^{-1} or more. This means that the interstellar gas must be associated with a fairly bright star— preferably of O or B spectral type—which provides the energy necessary for an optical transition. On the other hand, any cloud through which a bright star is clearly visible probably does not have the high opacity (in this case, grains along the line of sight) required to produce significant numbers of complex molecules or to protect from photodissociation those few molecules produced. In a comet, parent molecules in a region of high opacity receive more than the necessary excitation energy in the form of dissociating ultraviolet radiation from the Sun; the resulting molecular fragments produce fluorescent spectra that are easily observed. Hence, optical astronomers have detected many more cometary molecules (such as CH, CH$^+$, CN, C$_2$, C$_3$, NH, NH$_2$, N$_2$$^+$, CO$^+$, and OH) than interstellar molecules.

Unfortunately, in the past many radio astronomers did not recognize several basic differences between optical and radio observations. Gas and dust attenuate radio waves much less than optical waves; thus, molecular clouds observed by radio techniques may have much higher opacities than their optical counterparts. The greater number of interstellar grains along the line of sight is then responsible for the higher observed abundances of complex molecules. In addition, the molecular excitation conditions required for radio detections are not as stringent as those for interstellar and cometary observations of optical transitions. Pure rotational transitions are commonly found by radio techniques; the typical excitation energies required are below 100 cm^{-1}. Such small amounts of energy may be easily obtained from collisions with neutral particles traveling at thermal speeds or even from the general background microwave radiation. Hence, radio detections do not require particularly special geometrical placements, such as a bright star with intervening molecular clouds of low opacity.

What about receiver technology? Many of the new molecules (such as CO, HCN, and HC$_3$N) required new developments in receiver technology before they could be detected. On the other hand, most of the masering interstellar H$_2$O signals are so intense that they probably could have been found in 1963, shortly after astronomers realized that the interstellar clouds contain OH. The 4830-MHz line of H$_2$CO was detected on a receiver which had been used to observe the H 109α recombination line of atomic hydrogen at 5009 MHz. It is not unreasonable to say that astronomers are currently detecting some molecules that could have been found at least eight years ago, and the main reason for this delay was a general reluctance to consider the existence of polyatomic molecules in the interstellar clouds.

Let us hope that the same type of nihilistic logic that delayed progress in astronomical molecular spectroscopy for years will not be invoked again to try to halt or slow down the rate of new molecular detections. The chemistry and thermodynamic structure of the interstellar clouds will begin to be understood only when the complete molecular environment has been fairly sampled.

5. Life in the Universe

Franz D. Kahn

University of Manchester

The fundamental particles that constitute the material of the Universe are the proton, the electron, and the neutron. Their properties are

	Mass	*Charge*	*Spin*
proton	1.6×10^{-24} gm	$+e$	$1/2$
electron	9×10^{-28} gm	$-e$	$1/2$
neutron	1.6×10^{-24} gm	0	$1/2$

All three particles are extremely light; so much so that even the smallest living cell will contain 10^{12} atoms. The Earth itself contains a much larger number, of the order of 10^{51}, while a typical star like the Sun contains 10^{56} or 10^{57} atoms. The exact figures do not matter so much to us here; we are much more concerned with noting that stars, planets, and living systems are all physical objects made up of very many separate particles. The crucial difference between such systems lies not in just how many particles each system contains, but in the manner in which it is organized. This essentially is our concern in this essay.

The charges on the electron and on the proton are extremely important in determining the nature of their interaction. The quantity e, in terms of which each charge is defined, is also rather small. A current of one ampere, such as might be used in an electric light bulb, is equivalent to the passage along the filament of the bulb of about 6×10^{18} electronic charges per second. Finally, the spin of an electron, a proton, or a neutron is one half of $\hbar$, the basic unit spin. The spin of a single electron, proton, or neutron can be positive or negative, but it must have the numerical value $\hbar/2$. Again $\hbar$ is a small quantity; the wheel of an automobile, moving at 30 mph, has a spin of the order of $10^{36}\,\hbar$. These examples seem to show that the microscopic nature of matter is not very relevant to our everyday concerns; after all, who could possibly notice the flickering of a light due to the passage of all those individual charges through the light bulb, or the 10^{36} separate jolts as a car is brought to rest. However, we say it again: the very properties of the materials with which we are familiar depend in an intimate way on the properties of the subatomic particles of which they are made.

By this argument we could, if we wished, establish a connection between any two branches of human activity, say the study of elephants and space research. Does not this present attempt, to link astrophysics with biophysics, fall into the same category of irrelevance? I think the answer is definitely no. There are many important links. For example, the chemistry of living systems is much concerned with the properties of compounds involving carbon, nitrogen, and oxygen. These elements do not exist everywhere in the Universe, and they only come to be formed under rather special conditions in highly evolved stars. Again, living systems inevitably depend on existing in surroundings that are not in thermodynamic equilibrium. No ecology would be possible for a system placed inside a completely insulated enclosure: in our case the necessary disequilibrium is caused by radiation from the Sun. Without sunlight there would be no life on Earth. Now astrophysics tells us why stars must form and why they must shine once they have formed. It also tells us how long and how intensely the stars will radiate. Clearly these are valuable data in any search for locations that might be suitable for living systems. Now let us look in some greater detail at the questions we have raised.

Interactions

There are four basic kinds of force between material particles that are important to us. They are due to gravitational, electromagnetic, nuclear, and weak interactions.

1. *Gravitation* is the most familiar of these. It always attracts masses toward one another and becomes important when at least one of the masses involved is large. Thus we readily notice the Earth's gravitational pull on us but ignore the gravitational pull of smaller, everyday objects. Gravitational forces hold stellar systems together. For example, stars are made to travel in orbits about the center of a galaxy because of the gravitational pull that they experience. The planets in the Solar System travel in orbit around the Sun because of the Sun's gravitational attraction. Finally, the Sun itself is held together by its self-gravitation. But it does not collapse inward onto itself because its constituent particles are all in (thermal) motion.

2. *Electromagnetic forces* cause the attraction of unlike charges and repulsion between like charges. Thus an electron is attracted by a proton; in part this effect leads to the formation of a hydrogen atom. But we need to know more before we can describe the atom, and defer this for the moment. Both the gravitational and the electromagnetic interactions are long-range; that is, they extend over large distances even though they fall off in intensity as the particles move farther apart.

3. But *nuclear forces* are short-range, and they attract nucleons (*i.e.*, protons, neutrons) toward each other. Here short means very short indeed. The range of nuclear forces is of the order of 10^{-13} centimeter; within this distance they are very powerful. Nuclear forces can bind together protons and neutrons into the different nuclei: the properties of some important nuclei are

Nucleus	*Symbol*	*In terms of protons and neutrons*
Helium	$_2\text{He}^4$	2p + 2n
Carbon	$_6\text{C}^{12}$	6p + 6n
Nitrogen	$_7\text{N}^{14}$	7p + 7n
Oxygen	$_8\text{O}^{16}$	8p + 8n
Iron	$_{26}\text{Fe}^{56}$	26p + 30n

Nuclei are very small, of the order of 10^{-13} centimeter in linear dimensions. In the typical nucleus there are approximately equal numbers of protons and neutrons. At short enough range, nuclear forces are powerful enough to overcome electrostatic repulsion, so that many protons can be bound together into a nucleus. But once again our description is not complete; as in the case of the atom we must still explain why nuclear forces cannot pull the nucleus together into a singular point.

4. The *weak interaction* binds the proton to the electron to form a neutron. But the neutron is weakly unstable and will decay, in free space, within ten minutes to a proton and electron. When the neutron is bound into a nucleus it is surrounded by many protons, and is then usually stable. But if a neutron is not stable when bound into a nucleus, then the nucleus concerned is radioactive. For example, in the radioactive series beginning with uranium ($_{92}\text{U}^{238}$) there are several such steps, including the decay of radium A ($_{84}\text{Po}^{214}$), radium B ($_{82}\text{Pb}^{214}$), radium C ($_{83}\text{Bi}^{214}$), and so on. In each case a nucleus containing an overabundance of neutrons emits an electron and partially corrects for the neutron excess.

Two Important Restrictions

The Heisenberg uncertainty principle and the Pauli exclusion principle place two important restrictions on the possible interactions between protons, electrons, and neutrons and, indeed, between all microscopic particles. The Heisenberg principle states that momentum and position cannot both be precisely defined for any one particle at the same time. In any real description of a particle both are only approximately known, the momentum to within an accuracy δp, the position to within an accuracy δx, and the product $\delta p \delta x$ must always exceed $\hbar/2$. The Heisenberg principle applies to all particles.

The Pauli exclusion principle, however, applies only to *fermions* (so called after Fermi, the physicist), that is, to particles with half-integral spin. Protons, neutrons, and electrons belong to this class. The principle requires that only two fermions can occupy an elementary cell in phase space. To do even this much the two particles must have oppositely directed spins. Phase space, in this definition, is represented in six dimensions and is used to describe the momentum components of a particle as well as its position coordinates. In accordance with the Heisenberg principle, an elementary cell in phase space has the six-dimensional volume $(2\pi\hbar)^3$: each of the three pairs of sides of the six-dimensional cell must have a minimum area $2\pi\hbar$. This is the least possible value for the product of a step Δp in momentum and a step Δx in position. One of the important consequences of the Pauli principle is that an electron cannot be fitted into a small space unless enough energy is available to supply the electron with the necessary momentum. Diagram 5.1 indicates how the energies compare in the case of the hydrogen atom.

Diagram 5.1 gives the minimum radius that a hydrogen atom can have. The radius is about 10^{-8} centimeter for an atom in its lowest, most strongly bound state. This is the ground state of the H atom. It has a binding energy of 13.6 electron volts.

The hydrogen atom is simple because the nucleus has charge e, and the atom therefore holds just one electron, with charge $-e$. Other atoms have different charges on their nuclei: for example, the carbon atom has a nuclear charge $6e$ and therefore holds six electrons. These six electrons arrange themselves around the nucleus in a more complex pattern, which nevertheless must respect the Pauli principle. Calculation shows that two electrons fill the innermost shell. By application of the Pauli principle it is found that the next shell can hold up to eight electrons, the third shell may hold up to

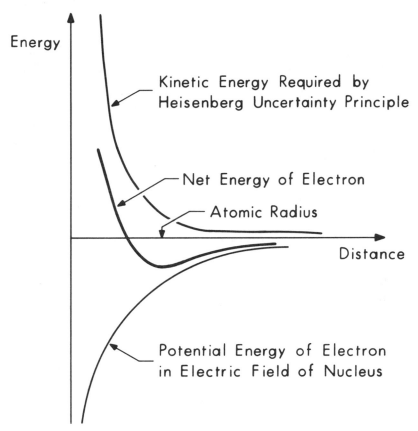

Diagram 5.1. The atomic binding curve (*center*) for an electron that is electrostatically attracted to an atomic nucleus, while at the same time moving in accordance with the Heisenberg uncertainty principle. The finite atomic radius corresponds to the minimum point on the net energy curve.

eighteen, and so on. In the carbon atom there are therefore six electrons, two in the inner and four in the outer shell. This leaves four vacant states there. The nitrogen atom, with seven electrons, has three vacant states in the outer shell, and the oxygen atom, with eight electrons, has two such vacancies. In the language of the chemist, the valencies of carbon, nitrogen, and oxygen are four, three, and two, respectively. But argon atoms, with ten electrons, and helium atoms, with two electrons, both have completed outer shells. Now chemical reactions are determined by the interplay of forces between the electrons in the outer shells of atoms. If the shells are incomplete the atoms readily enter into chemical reactions, but if the shells are complete, as in the case of helium and argon, then the element is inert. By considering some simple well-known compounds we can see how the atoms come together in such a way as to fill the outer shells. For example:

Compound	Constitutents	Outer shell electrons	Total
Water	H + H + O	1 + 1 + 6	8
Ammonia	H + H + H + N	1 + 1 + 1 + 5	8
Carbon dioxide	C + O + O	4 + 6 + 6	8 + 8

In each case the electrons try to rearrange themselves in completed shells. Energy is released in the process, just as it is released when an outer electron is bound into an atom. These interactions form chemical (covalent) bonds between atoms and have binding energies of a few electron volts, the characteristic binding energy for an outer electron in an atom. Under typical conditions in a star there is usually enough thermal energy available to break all such bonds; therefore, we do not find chemical compounds in any but the coolest stars. But on Earth the temperature is much lower, and therefore chemical compounds are very abundant. There are also chemical compounds present in interstellar space. First, the interstellar dust probably consists of frozen solid material, like H_2O. Second, various molecules have been observed directly, largely by radioastronomical means. The common view is that these molecules are formed on the surfaces of the dust grains: some of the molecules, like wood alcohol and formaldehyde, look distinctly organic.

Now let us reconsider the structure of the atomic nucleus. A proton or neutron is bound into the nucleus by short-range internuclear forces. The exact form taken by such a field of force is not known for small interparticle distances much less than say 10^{-13} centimeter. We shall accordingly represent the variation of the potential energy by the squared-off curve shown in Diagram 5.2. In the same diagram we also show the minimum kinetic energy that the nucleon needs in order to be confined to the small volume of the nucleus. The two cases A and B are rather different. In case A (Diagram 5.2), the particle will certainly be caught in a trough of negative energy. It is thus firmly bound to the nucleus. The α-particle ($_2He^4$) is a good example of a firmly bound nucleus. Energy is released when nucleons are captured into such a nucleus. We call a capture like this *exothermic*. Such reactions are the source of nuclear energy in stars.

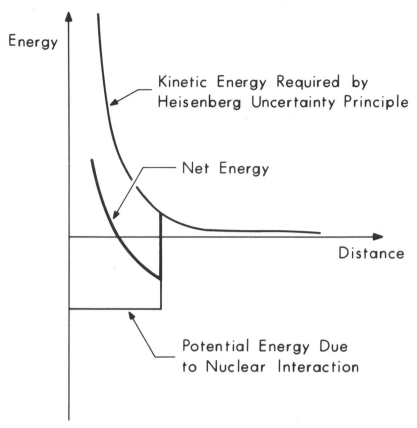

Diagram 5.2. Stably bound nucleons in an atomic nucleus result when the net energy curve is negative at its minimum. The strong nuclear force overcomes both the electrostatic repulsion of the nuclear protons and the Heisenberg energy.

It is, incidentally, clear that the Heisenberg uncertainty principle ensures that a nucleus has a finite radius, since the finite amount of binding energy available restricts the momentum of any nucleon.

Now in case B (Diagram 5.3), the nucleon is again trapped in an energy trough, but this time the net energy stays positive. Therefore the process of formation of such a nucleus is not exothermic, but *endothermic*; that is, energy must be supplied to allow it to form. In fact the trapped nucleon (or frequently the trapped α-particle) can leak its way out through the potential barrier, with considerable release of energy. The radioactive uranium series contains many examples of such (α-particle) decays, for instance, those of uranium I ($_{92}U^{238}$), uranium II ($_{92}U^{234}$), ionium ($_{90}Th^{230}$), and so on. The

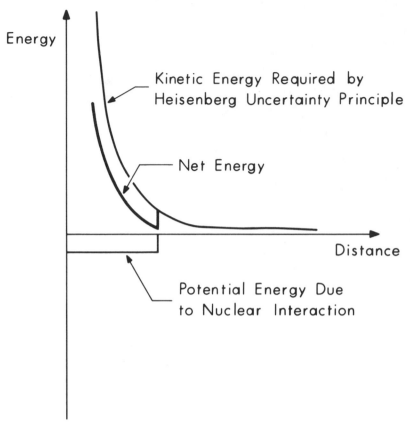

Diagram 5.3. Unstably bound nucleons in an atomic nucleus can "tunnel" through the energy barrier to freedom (radioactivity). Here the net nuclear energy trough is so shallow that it is always at positive energies.

half-lives of these nuclei are respectively 4.5×10^9 years, 2.5×10^5 years, and 8×10^4 years. The half-life of radium ($_{88}Ra^{226}$), which also decays in this way, is only 1620 years. In general, therefore, the nucleus of an atom has a permanent existence only if it was formed by an exothermic process.

The Existence of Structures

The interatomic forces can give rise to various structures with well-defined geometric features. We shall illustrate our argument in terms of the structure of a metal. But qualitatively the same argument can be used to consider the geometric features in the structure of a water molecule, or an organic molecule, like a benzene ring, or a macro-molecule like DNA or even a protein. In all cases the structure persists, in the last resort, owing to the great difference between the mass of an electron and the mass of an atomic nucleus.

Take then the case of a metal, say iron. The outer (valence) electrons of each atom enter into the interaction that binds the metal together; with two valence electrons per atom this leaves the atom ionized, with charge $2e$. This Fe^{++} ion is now held in position by the electron cloud that surrounds it. If the ion departs too far from its equilibrium position, it experiences a force tending to pull it back. Calculation shows that the mean square momentum $\langle p^2 \rangle$ of the ion is related to the mean square displacement $\langle r^2 \rangle$ by the order of magnitude relation

$$\langle p^2 \rangle \sim \frac{8\pi}{3} N m_A V e^2 \langle r^2 \rangle$$

for a substance with valency V. But the uncertainty principle tells us that in the lowest energy state of a system

$$\langle p^2 \rangle \, \langle r^2 \rangle \sim \frac{\hbar^2}{4},$$

and so we deduce that

$$\langle r^2 \rangle \sim \left(\frac{\hbar}{2e} \right) \left(\frac{3}{8\pi NV} \right)^{1/2} m_A^{-1/2}.$$

Thus the mean square uncertainty in the position of the atom varies inversely as the square root of m_A, the mass of the atom. We can in the same way work out $\langle R^2 \rangle$, the mean square uncertainty in the position of an electron, and find, by a reasonably analogous argument, that

$$\langle R^2 \rangle \sim \left(\frac{\hbar}{2e} \right) \left(\frac{3}{8\pi NV} \right)^{1/2} m_e^{-1/2}$$

where m_e is the mass of the electron. From these two results we have

$$\frac{\sqrt{\langle r^2 \rangle}}{\sqrt{\langle R^2 \rangle}} \sim \left(\frac{m_e}{m_A} \right)^{1/4}. \tag{1}$$

Now $\sqrt{\langle R^2 \rangle}$ is about equal to the interatomic distance. The atoms cannot get closer than this, since otherwise the electrons would be forced into a smaller space. Therefore our result (1) states that the uncertainty in the position of an atom in the structure is about $(m_e/m_A)^{1/4}$ times the interatomic distance. The factor is of order 1/10 for most substances. The relative position of two atoms in the structure is therefore defined with about 10 percent precision. In a solid there are many links between the atoms, and the resultant structure is quite rigid. A rather similar argument could be used to establish that an organic molecule has a well-defined structure. But the structure will fail if many of its atoms cannot be located with the precision required. This will happen when the substance becomes too warm, so that the oscillations of the atoms allow them to move too far from their mean positions. We then say that the metal melts, or that the molecules dissociate. The continued existence of DNA, or a protein, or an enzyme is assured only if the temperature does not rise too high.

More fundamentally, we shall now argue that the existence of structures with a well-defined geometry is made possible exclusively by chemical forces. Only with chemical forces do we have interactions between particles whose masses differ by a large factor: the ions and the electrons. The Pauli exclusion principle ensures that the nuclei stay a respectful distance apart. The Heisenberg uncertainty principle then shows that the positions of the atoms (or ions) can be defined with good relative precision, provided the temperature is not too high.

But suppose that we were to attempt to construct a symmetrically ordered system, based on the gravitational or the nuclear interaction. In the first case the Heisenberg and Pauli principles impose no real restriction since the "particles" we are dealing with—that is, stars or planets—are extremely massive. Only dynamical equilibriums, with fluid patterns, are therefore possible. In the second case both principles can sensibly be used, but the interactions concerned take place between protons and neutrons, that is, between particles of equal mass. No precision can then be given to the position of any nucleon. From the outside all nuclei look spherically symmetric.

Leaving aside the weak interactions, which seem to be decidedly unpromising for our present purpose, we are thus driven to conclude that geometrically ordered systems can exist only in conditions where chemical forces are important. Living systems are necessarily ordered.

Energy Considerations

Enough of all this physics; let us now get down to some applications. Our first and most important point is: there is unlikely to be complete equilibrium in a system where gravitation is important. This is the case in astronomical objects. As is well known the Universe expands. In the beginning its material consisted entirely of hydrogen and helium. But later condensations developed, as a result of gravitational contraction, and many of these can now be seen as galaxies. Within any galaxy, say in our own, the gas eventually became sufficiently cool and quiescent that further contraction followed eventually to form stars.

A star, once formed and settled down, constitutes a system with negative energy. It can rigorously be shown that the total energy E, which is the sum of the internal energy U and the gravitational self-energy V, is always negative, and that $U = -E$. But U is proportional to the temperatures in the star. Therefore, as the star radiates energy into space and as its energy becomes more negative, so its internal energy U and its temperature T both increase. The star therefore be-

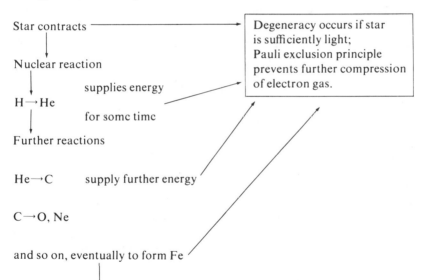

Diagram 5.4. A schematic outline of stellar evolution for stars of low and high mass. Interstellar hydrogen is transmuted to heavier elements in this process, and may enrich the interstellar medium.

comes progressively hotter inside. The possibilities shown in Diagram 5.4 then arise.

We therefore conclude, since the stars fill only a small fraction of space, and since they are hot, that the conditions in the space around any star are very far from thermodynamic equilibrium. For at any point, the starlight comes only from a very small portion of sky, which is bright. The rest of the sky is dark and cold. This imbalance leads to a number of well-known nonequilibrium effects, particularly in interstellar space. There the energy density of the radiation corresponds to that in a black enclosure at about $3°K$, but the gas temperature ranges between $50°K$ and $10,000°K$, and the temperature of the solid grains in space is about $10°K$. There are even components of the gas to which no sensible temperature can be ascribed. Examples are given by the regions containing the OH masers and the formaldehyde inversemasers.

From the present point of view the nonequilibrium state of the interstellar dust grains is particularly interesting, since it is widely believed that the grains provide the site for the formation of molecules in interstellar space. Conditions could hardly be further from equilibrium: to describe them a separate temperature must be assigned (if possible) to the color of the radiation field, to the energy density of the radiation field, to the interstellar atoms that hit the grain, and to the lattice vibrations of the grain. Finally the situation is further confused by the presence of cosmic ray particles, whose average energy is very high indeed. Needless to say, the surface chemistry of interstellar grains is not well understood. But observation indicates that many molecules can readily form on grain surfaces. Examples are CO, CH, CN, OH, H_2O, NH_3, HCN (hydrogen cyanide), HC_3N (cyanoacetylene), H_2CO (formaldehyde), and CH_3OH (methanol). Everyone expects that many more molecules will be discovered soon.

The solar radiation field, as seen at the Earth, is also rather dilute, though not as dilute as the radiation field in interstellar space. The Earth is illuminated by the Sun, over about 0.0005 percent of the sky, with a temperature of $6000°K$. This provides enough energy to maintain the terrestrial equilibrium temperature at about $300°K$. But sunlight contains photons whose energy is much higher than would normally be found at $300°K$. These photons can therefore initiate chemical reactions (*e.g.*, photosynthesis) that would normally be impossible at $300°K$.

Now we shall briefly consider another consequence of stellar evolution. The nuclear processes within the star lead to the formation of progressively heavier nuclei. We know that stars of the earlier generation (Population II) were composed almost entirely of hydrogen and helium. Only at a later stage in the evolution of the Galaxy do we find heavier elements like C, N, O, Si, Fe, and so on. Therefore the raw material for the building of planets, as for the genesis of living systems, only became available after some stars had evolved, and had ejected back into space the transmuted material in their interiors. Before that time there was not much scope: one cannot do much interesting chemistry with just hydrogen and helium. It is also rather hard to form a planet out of these elements unless one puts the planet in a chilly place. But once the elements silicon, iron, and oxygen are available, then solid substances exist at more reasonable temperatures, and the formation of planets near a star becomes much easier. With the elements carbon, nitrogen, and oxygen available the possible repertoire of chemical reactions becomes much richer and more promising. According to our present ideas the formation of the Earth, and of the other planets, took place at the time when the Sun itself was formed. The Sun is a star belonging to the later generation (Population I); at the time of its birth the heavier elements, like C, N, O, Si, and Fe, were available in more or less their present-day abundance. Soon after the formation of the Earth there began the chemical reactions that we believe to have led to terrestrial life. It is usually thought that these primitive events took place in the sea, the so-called *primeval soup*. The necessary chemical compounds, like, for example, Adenine, Cytosine, Guanine, Thymine (as in DNA), and many others, are thought to have been formed after the heating of the atmosphere by meteor falls, volcanic eruptions, or lightning flashes. These compounds would then be washed into the sea and remain dissolved there. But there is an alternative: it is entirely possible that a sufficient abundance of organic material was brought down to the newly formed Earth along with the interstellar dust out of which this planet was made.

Necessities of Life

I shall consider just three necessities of life; they are tradition, food, and sunshine.

First, *tradition* is intended to be understood in the sense that any living organism is itself descended from a previous generation. The whole species to which it belongs will have evolved over many generations to fit in with its surroundings and with the company it keeps. We know that the information is contained in the genes of the living system, and that the genetic code is inscribed by the order of occurrence of the bases Adenine, Cytosine, Guanine, Thymine, in the DNA molecule. This coding is, of course, highly geometrical. One of the essential features of the code is the manner in which the DNA molecule replicates, so that copies may be made of the genetic information. As is easily seen from any model of its structure, the DNA molecule has the shape of the well-known double helix: at regular intervals inside the double helix there are pairs of bases. The bases are always paired off in the same way. This happens because given this group of bases the selection and arrangement of the atoms in the inner side of the helix is such that A only fits onto T and G

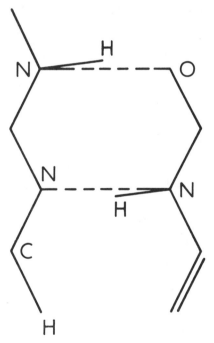

Diagram 5.5. The A-T base-pair in the DNA double helix consists of an Adenine molecule (*left*, with two nitrogen atoms shown) bound to a Thymine molecule (*right*, with oxygen and nitrogen) by two hydrogen bonds (dashed).

only fits onto C, and vice versa: two hydrogen bonds being made in one case, and three in the other (Diagrams 5.5, 5.6, and 5.7). The H bonds are weaker than covalent bonds. They are essentially made by electrostatic effects: the H atom polarizes, so do its neighbors. Replication of a DNA molecule occurs when the two strands of the helix separate: the bases on each separate strand then acquire a new set of mates; these are later zipped up by the phosphate molecules that form the backbone of the helix, and two DNA double helices have grown where only one existed before. Clearly the process of transmitting genetic information is itself based very closely on the transmission of data by geometrical means, *i.e.*, the location of the A, C, G, T molecules in the chain. It seems that any living system must incorporate this basic feature; perhaps other living systems are not based on DNA, but they must surely be based on some organic macromolecule with analogous properties.

Next, *food*. The living system must be doing something. If it is not

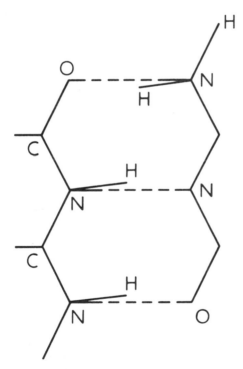

Diagram 5.6. The C-G base-pair in DNA consists of Cytosine (*right*) joined to Guanine (*left*) by three hydrogen bonds (dashed).

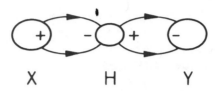

X   H   Y

Diagram 5.7. A weak hydrogen bond arises when a
hydrogen atom (*H* at *center*) polarizes (− +) two
neighboring molecular atoms (*X* and *Y*).

doing anything, we say that it is dead. To do anything requires energy, and the living system extracts this energy from its food. Therefore food, in its most general sense, is a substance in which a set of
particles are held together in a certain way, but in which a rearrange-
ment of these particles leads to a state in which the same set of particles has a lower potential energy. The difference in potential energy is made available to the system and allows it to do the work that
it has to do. In this general sense we could think of food in a variety
of ways.

"Food"	Method of energy release	Comments
1. Radium	Radioactivity	Energy can be stored a long time; but no control exists over time of release.
2. Hydrogen (as in the H bomb or in a star)	Raising temperature	Energy can be stored; release possible only under extreme conditions.
3. Wood	Lighting of a fire	Energy can be stored; but heat must be supplied to release it.
4. Bread	Digestion	Energy can be stored and released in specific places and at specific times by specific chemical reactions, catalyzed by enzymes.

The difference between cases 1, 2, 3, and 4 is clear. In cases 2 and
3 the energy can be released at will, but only by sledgehammer

methods. In the case of a star or a wood stove we do not care too much about this: neither the star nor the stove has a delicate structure which needs preservation. Of course, anyone who uses an H bomb is out to do as much damage as he can; so he welcomes this feature of the process. Case 1 has the disadvantage that there is no control over the process of energy release, but it does occur without any need to raise the temperature. Case 4 can occur without raising the temperature and under controlled conditions. In a way the action of an enzyme is like the action of a syphon: we may use a syphon to empty a tank of water without having to do the work of lifting the water over the top of the tank. In the biochemical reaction the enzyme can act at particular stages in the chemical process because a particular part of its own configuration matches that of the substances that are being processed, and the presence of the enzyme in the correct place allows a potential barrier to be depressed. Further, it becomes possible for a certain sequence of reactions to occur in a certain order if the various necessary enzymes are arranged in a certain order.

Now the reason why things must occur in this way is simple. The food that the living system requires should not decompose of its own accord; therefore, there must be a potential barrier which prevents its decomposition (just as there is a potential barrier for the α-particles trapped in a radioactive nucleus). In the case of the radioactive nucleus, the α-particle eventually leaks through the barrier of its own accord; in the case of the wood stove, the potential barrier is broken when the temperature is raised sufficiently high. Neither of these methods of energy release is suitable for a living system; in particular, by raising the temperature to release the energy in the food, we would also destroy the very structure of the organism. For this reason it seems that any living system needs a suitable set of catalysts (enzymes) to decompose its food, especially if various parts of the system require different energy inputs at different times.

Finally, *sunshine*. How did the various particles come together in a metastable configuration which has a considerable store of potential energy? In the last resort this is due to the action of some outside source of energy, whose characteristic temperature is much above that of the living system concerned. In our case the ultimate source of energy is the Sun, at a temperature of $6000°K$. The great temperature difference between the Sun and the Earth arises because the sunlight that heats the Earth comes from only a small fraction of the sky, but the Earth reradiates the heat to the whole sky, and the sky is cold. The Earth and its atmosphere are therefore necessarily cooler than the Sun. As far as we can tell the Sun will slowly get more luminous as it evolves, and the sky will slowly get cooler as the Universe expands. Therefore the temperature differences will remain, at least until the Sun stops shining. Hopefully we may expect that life will continue until that time.

Where Should We Prospect for Other Living Systems?

It is possible to make a few further guesses to help narrow down the possible range of those places where life might exist.

1. There has to be a process of arranging the various organic molecules into a collection of macromolecules, which can then function as a living system. This is likely to take a long time and to require many trials and errors. Therefore we expect the chances of success to be largest when there is a primeval soup, with a good concentration of the necessary organic compounds. Further, the substances must be well mixed and continually stirred; so that the best chance occurs if the soup is liquid. It seems that water is a very promising solvent for the purpose. Other possible solvents exist, but water seems the most likely, since the abundances of the elements in Population I suggest that H_2O should be one of the most abundant compounds. If we consider life based on water, then we must seek a place for our living systems where water is liquid: again, neither too near nor too far from a star.

2. Many molecules are needed, and there must be many interactions between them, in order for them to constitute a life system. Therefore we would not expect such a system to organize itself in interstellar space. The concentration of molecules in space is so low that encounters between molecules would be very rare indeed. Only in the layers of molecules adsorbed onto the grain surfaces would the molecular density be high. But at the best of times a grain can hold only some ten or a hundred million atoms adsorbed at its surface. This gives little scope for interesting developments.

3. The life system needs time to evolve, and presumably it has a better chance of doing so if conditions remain reasonably stable. Therefore we should probably exclude from our search the planetary systems of massive stars, for these stars evolve too fast. Further, climatic conditions are likely to be rather irregular on planets be-

longing to binary or multiple stars, whose orbits are much less likely to be simple and circular. Finally, if a star is not massive enough then it will also be rather dim, and a planet would have to be improbably close to the parent star if it is to be warm enough for water to be liquid.

To summarize, then: our best chance of finding a living system is to look among the planets of those single stars belonging to Population I, whose masses are within 50 percent or so of the mass of the Sun. Have we now unduly restricted the range of possible places? A rough estimate would suggest that there are about a billion (10^9) such stars in this Galaxy alone, and it seems that there are some 10^{11} galaxies in the Universe. We conclude therefore that while life can only arise under a rather restricted range of physical conditions there are very many places in the Universe where those conditions are fulfilled. If we want an answer, we should encourage a biophysicist to tell us how likely it is that life will develop if the chemical and climatic conditions are suitable.

6. Galaxies: Landmarks of the Universe

Morton S. Roberts

National Radio Astronomy Observatory

Today's city dweller is denied one of the striking scenes of Nature: a clear view of the night sky. Atmospheric pollution and the glow of lights drown out all but the Moon and the brightest stars and planets. This is in striking contrast to the appearance of the sky far from the city environment. Here, on a clear, moonless night the naked eye easily sees a thousand stars. Every few minutes a meteor flashes across the sky. Laced around the celestial sphere, appearing like thin clouds, is a band of light aptly named the Milky Way.

Historical Context

The first correct explanation of the appearance of the Milky Way was given by the Greek philosopher Democritus (born 460 B.C.), who maintained that "it was composed of a multitude of small stars, so very near to each other, that their light blended together so as to produce the appearance of a luminous zone" (see Fig. 11). It was not until 2000 years later that Galileo's observations showed this explanation to be correct. Galileo described these very first astronomical observations made with a telescope in a small booklet, *Messenger of the Stars*, published in 1610 A.D. As a result, attention was refocused on the Copernican doctrine wherein the Earth, and now the Sun, lost its privileged central position in space, and a new era began in our concept of the structure of the Universe.

The next major step lay 300 years ahead. Extensive observations, cataloging, and measurements complemented by theory, calculation, and speculation prepared the groundwork. By 1920 it was recognized that our Solar System is located toward the outer edge of the disk of stars that make up the Milky Way Galaxy. With more recent work, the details of this scale have become clear: we are located some

The National Radio Astronomy Observatory is operated by Associated Universities, Inc., under contract with the National Science Foundation.

30,000 light-years from the center of the Milky Way, a flat stellar system 100,000 light-years in diameter. The overriding astronomical question of the first quarter of this century—answered in 1923—was: Is the Universe made up of just our Milky Way Galaxy or are there other galaxies scattered throughout space? The question itself was not new. As early as 1750 Thomas Wright, and five years later Immanuel Kant, had proposed that our Universe is populated with many galaxies, each of a size comparable to our own Galaxy.

These suggestions were, in part, based on the many nebulae that astronomers had cataloged.* There is an obvious parallel with the development of our view of the Milky Way. A distant galaxy could not be resolved into its member stars with the small telescopes first available, just as the Milky Way could not be resolved into individual stars with the naked eye. Although Wright and Kant offered explanations for the nebulae, their theories were little more than speculation, since no distances to these nebulae were available. Distance was the key yet to come, and lacking it the theory of galaxies beyond our own fell into disfavor.

The advent of spectroscopic astronomical observations in the nineteenth century rekindled (and complicated) the argument but did not solve it. An important clue, which became meaningful only after the independent establishment of the extragalactic nature of many of these nebulae, was the large displacement of their spectral lines, relative to the wavelengths of the same lines in the laboratory. Interpreted as Doppler shifts, these displacements gave velocities as high as 1100 kilometers per second, much larger than any stellar radial velocity. Fifteen nebulae had been studied by V. M. Slipher by 1915. Of these, eleven showed velocities of recession. The observational basis for the concept of an expanding Universe was at hand.

On October 5, 1923, Edwin Hubble obtained a photograph of the Andromeda nebula (Fig. 12) with the 100-inch telescope at Mt. Wilson in California. He was studying various features of the nebula, and this plate was one of a series he had taken. He later entered in his observing log for this date the following note: "On this plate, three stars were found, two of which were novae and one proved to be a variable later identified as a Cepheid—the first to be recognized

*Nebula is Latin for "cloud." These hazy patches of light were seen through telescopes over the entire sky, and in 1781 Charles Messier cataloged 103 such objects so that other astronomers would not mistake them for comets. Some nebulae are gas clouds or clusters of stars in our Galaxy, but most are extragalactic stellar systems.

in M 31 [Andromeda]." With these simple words, Hubble had solved the problem; there were galaxies other than our own!

The Cepheid that Hubble found was the yardstick with which to measure the distance to the Andromeda galaxy. The importance of finding a Cepheid variable star lay in the fact that the rate of light variation of such a star is related to its intrinsic luminosity.* From a series of observations, the period of variation and the average apparent luminosity are derived. The distance of the star, and hence of the galaxy, follows directly from the inverse-square relation between the intrinsic and apparent luminosities. The most recent determination of this distance places the Andromeda galaxy two million light-years away, clearly beyond the confines of our own Milky Way Galaxy. Since the observed angular extent of Andromeda on the sky is ten to a hundred times greater than that of many other bright nebulae, it was reasonable to assume correspondingly larger distances for these other nebulae. This assumption has now become a well-established fact.

Half a century has passed since Hubble opened the door to the other galaxies. Progress in extragalactic research has been difficult and slow. Photographs and spectra used to require hours, even days, of exposure time; now the same job can be done in minutes using image intensifiers. The galaxies are slowly yielding their secrets, as we extend our probes throughout the electromagnetic spectrum from γ-rays to radio waves. Since World War II, extragalactic surprises have multiplied profusely: the Hubble expansion has been verified more than halfway to the "edge" of the Universe, wondrously strange galaxies have been uncovered (peculiars, Seyferts, exploding galaxies, and radio galaxies), and the enigmatic quasars (quasi-stellar sources) have burst upon the astronomical scene only within the past decade.

What do we think that we know about extragalactic objects, and where are our discoveries leading us? These are the questions that I will briefly attempt to answer in this essay.

What Is a Galaxy?

A *galaxy* is a gravitationally bound conglomerate of many stars; 10^6 stars is probably a good minimum number, for we do not want

---

*These bright giant stars pulsate with periods from 1 to 100 days. As they expand and contract, they rhythmically grow brighter and dimmer. The periods and average brightnesses of Cepheid variables are correlated by the period-luminosity relation, discovered by H. Leavitt in 1912.

to consider globular clusters (see Fig. 19), which constitute but one subsystem of a galaxy. Our own large Galaxy is an excellent example: it is a flattened rotating disk of about 10^{11} stars, with "spiral arms" in the disk, a large "central bulge," and a "halo" of old stars and globular clusters. The nearest major galaxy is the Andromeda galaxy (Fig. 12), whose distance is 20 times the diameter of our Galaxy; hence, the descriptive phrase "island universes" is not a wholly inappropriate way to speak of galaxies. About 5 to 10 percent of the mass of our Galaxy is in the form of gas and dust between the stars; in addition, the interstellar space is permeated by starlight, cosmic rays, and weak magnetic fields. In the Southern Hemisphere we can see the Large and Small Magellanic Clouds, two nearby (200,000 light-years) satellite galaxies of our own Galaxy. It appears that galaxies can range in size from about 10^6 to 10^{13} stars (*i.e.*, solar masses), and they exhibit a fascinating variety of morphological forms and contents. Let us now consider the so-called normal galaxies in detail.

Classifying "Normal" Galaxies

Although galaxies show a variety of structural forms, Hubble was able to devise a remarkably simple scheme for their classification (see Diagram 6.1). First, two principal classes are identified: *ellipticals* and *spirals*. Within each class we apply finer criteria. The elliptically shaped systems, denoted E, are ranked by their apparent ellipticity from 0 to 6. The E0's are circular in appearance, while the E6's are the most flattened, with an apparent axial ratio of 2.5 : 1. The true ellipticity for any particular system cannot be determined. From the range of apparent ellipticities it is obvious that these systems cannot all be spherical. But are they all spheriods of a fixed oblateness seen at different angles, or is there an intrinsic range in the degree of flattening? These alternatives may be tested by computing the distribution of ellipticity in both cases; the second situation matches the observations best.

The *spiral* galaxies are further subdivided into two parallel sequences: *ordinary* (S) and *barred* (SB), obvious references to the absence or presence of a barlike feature in the central part of the system. The criteria that locate a spiral along these sequences are the angle of opening of the arms and the relative size of the central region or nuclear bulge, a feature which, if isolated, would appear

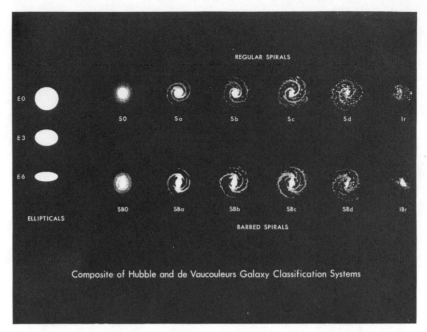

Diagram 6.1. Hubble's system of galaxy classification is shown in schematic form, with certain modern modifications proposed by G. de Vaucouleurs.

as an elliptical galaxy. These two criteria are correlated in the sense that tightly wound arms occur in systems with large nuclear regions, while open arms go with a small nucleus. An additional classification feature often invoked is the degree to which the arms can be resolved into individual bright stars and ionized hydrogen clouds. The degree of resolution correlates with position on the Hubble sequence. The stars and luminous clouds are most easily resolved in the open-armed, small-nucleus spirals.

At one extreme of this classification scheme are galaxies with arms so tightly wound and bulge so large that at first appearance the system resembles an elliptical; these are the S0's and SB0's. At the other end of this spiral sequence are systems with wide-open, highly resolved arms, in which the small nuclear region is all but lost in the arm structure. Hubble originally classified these as Sc (and SBc) systems. Later modifications introduced Sd and Sm (*m* for the prototype, the Large Magellanic Cloud), as even more extreme examples. In addition, there are galaxies described as *irregular*, which resemble a spiral arm section of an open-armed system. No nucleus is obvious. These irregular systems may be considered the extreme of the spiral classification. The notation for the sequence of ordinary spirals is: S0, Sa, Sb, Sc, Sd, Sm, Ir. A corresponding notation with the letter B is used for the barred spirals. Often, the Sm category is combined with the irregular systems. Diagram 6.1 shows a schematic presentation of the Hubble sequence, and four normal examples of different galaxy types are shown in Figure 21.

Although descriptive, this classification system is consistent and remarkably simple to master. Since these types are discrete, while the variation along a sequence and between sequences is continuous, the notation is easily expanded to accept intermediate classifications. For convenience, the phrase *early-type* is often used in reference to the S0, Sa end of the sequence, and *late-type* for the other (Ir) end; no temporal implications are intended by this phraseology.

Observational Selection Effects

A general correlation between total mass and galaxy type is only weakly suggested by the available data. Irregular and very late-type spirals have relatively low mass, 10^9 to 10^{10} solar masses. But late- and early-type spirals are not distinguished by mass; they cover the range of 10^{10} to 10^{12} solar masses. The most massive galaxies are ellipticals, reaching 10^{13}; but some of the lowest-mass galaxies known, 10^5–10^9, also belong to the elliptical category.* If there is a correlation it may exist for the maximum values of a wide distribution of values within each galaxy type. The difficulty in disentangling such information is due to the highly biased sample of galaxies that has been studied for mass determinations. Observational selection favors the apparently brightest systems that tend to be intrinsically bright and massive. Data for a complete sample of galaxies within a specific volume of space are not available.

Such observational selection is a common problem in astronomy and is particularly serious in extragalactic studies. Because of it we are unable to evaluate properly a variety of statistical data. An example is the true frequency (per unit volume of space) of the different galaxy types. A catalog, prepared by G. and A. de Vaucouleurs, of

*There is, apparently, some overlap with globular clusters (10^4–10^6 solar masses) at the low-mass end of the ellipticals. Are globular clusters always associated with large galaxies, or can they appear as "elliptical" interlopers in extragalactic space?

all galaxies brighter than apparent magnitude 13 has 7 percent of its spiral entries in the late categories of Sm and irregular. An estimated correction for selection effects raises this number to 33 percent; *i.e.*, one-third of all spiral galaxies are very late-type.

H. C. Arp has called attention to another type of observational selection that hampers a complete survey of extragalactic objects. We know of galaxies that have an almost stellar appearance and others of extremely low surface brightness, just detectable above the night sky brightness. Sky surveys will discriminate against these two classes of galaxies: the former because of confusion with stars in our Galaxy, and the latter because they are only marginally detectable from Earth-based observatories. (F. Zwicky is pioneering an effort to discover the stellar-appearing variety, to which he has given the descriptive name "compact galaxies"; he has cataloged several thousand thus far. In an attempt to discover extremely distant galaxies, A. Sandage is working at the sky-brightness limit of the 200-inch telescope at Mt. Palomar in California.)

The classification system is based on photographs taken with blue-sensitive emulsions. The photographic process is such that these emulsions are the fastest, and therefore they were the obvious ones to use for the early photographs of galaxies. With time, other fast emulsions were developed to be most sensitive in the yellow or red. When these were used a curious effect was found. This has been most clearly stated by E. Holmberg. He discovered "that the *majority* of spiral nebulae exhibit a more or less pronounced *bar*; the bar may not always be recognizable on the blue plate, but is usually visible on the [yellow] exposure. It seems quite possible that a bar is a structural detail common to all, or most, spiral nebulae." Thus the sequence of ordinary spirals might never have been introduced if yellow-sensitive, rather than blue-sensitive, emulsions had been used first. Here is a good example of the color telling us something about the stellar composition of the bar. Those spirals showing a bar structure only in yellow light must have a much higher proportion of yellow stars making up the bar than those whose bar is prominent in blue light. The barred sequence of galaxies is thus based on a more detailed criterion than just structural features; it contains information on the stellar composition of the bar.

Studying galaxies in the light of different colors offers great potential help in defining and understanding their structural features. With present data, the correlation of spiral class with the physical properties of integrated color and hydrogen content is the same for both ordinary and barred systems. It will be important to learn if this remains true after more detailed observations are made.

Correlations in the Hubble Sequence

The Hubble classification system is based only on the appearance of galaxies as seen projected on the sky. Does this apparently natural ordering reflect an ordering in the physical properties of galaxies? Does it also indicate some sort of evolutionary sequence? These questions are related, and we must answer the first before we can consider the second.

There are only two properties that clearly correlate with the structural sequence. These are: (1) the integrated color; later-type galaxies are bluer; and (2) the relative amount of neutral atomic hydrogen; later-type systems have more of their mass in the form of neutral atomic hydrogen. Thus, there is a significance deeper than mere appearance to the manner in which the galaxies are ordered.

Six factors determine the *observed color* of a galaxy: (1) the stellar content, (2) the dust content, which through its selective *extinction** properties will redden the system, (3) the emission lines from ionized hydrogen (HII) regions, (4) the orientation of the galaxy to our line of sight, which changes the path length through the system and hence the amount of reddening, (5) reddening due to the intervening dust in our own Galaxy, and (6) the red shift of the galaxy. The last three factors are not intrinsic to the galaxy. They can be evaluated and removed from the measured color. The resultant color is a measure of the stellar population as modified by the properties of the interstellar matter within the galaxy. Generally the dominant factor is the starlight, and the color thus enables us to gauge the mix of different types of stars within a galaxy. These stars are distinguished by their positions on the Hertzsprung-Russell diagram, as described in Essay 1.

The systematic variation of color with structural type can be understood in terms of a variation of the number ratios of stars of different spectral types. To derive a quantitative model of the stellar

---

*Interstellar "dust" absorbs and scatters starlight—the so-called *extinction*—and does so more strongly at shorter (bluer) wavelengths, leading to an apparent reddening of the light. The longer the light path through the dust (and the denser the dust), the dimmer and more reddened becomes the starlight.

population of a system requires additional information. The strengths of certain spectroscopic lines give luminosity criteria; that is, a distinction between stars of the same color but of different luminosities (*e.g.*, red giants and red dwarfs). A further constraint on any derived model is the total mass predicted by summing the masses of the individual stars in the model.

We find a very interesting result. The bluer colors of later-type systems show that they have a larger fraction of young, blue stars than the early-type galaxies. It is this prominent constituent of young stars that has led to the suggestion that late-type systems are in fact young. I shall return to this point later.

Now consider the *neutral atomic hydrogen (HI) content* of a galaxy, obtained by measuring the intensity of the 21-centimeter radio spectral line that such hydrogen emits. In the simplest case, the intensity of the line is directly proportional to the number of emitting atoms and inversely proportional to the square of their distance. This simple relation between the line intensity and the number of emitting atoms involves two assumptions. The first is that there are no absorption effects due to radiation sources lying behind the hydrogen. This assumption is valid for the vast majority of galaxies studied in 21-centimeter radiation. By analogy with our Galaxy, we do expect such radiation sources to be scattered throughout the plane of a galaxy. But the geometry of the situation is such that they can be neglected.

The second assumption is that the hydrogen is optically thin, that is, that a line of sight through a galaxy will encounter all the hydrogen atoms along the path. If opacity effects were present, the hydrogen at the end of the path would be "hidden" by hydrogen in front of it. Again, by analogy with our Galaxy we expect some hydrogen regions to have a high optical depth. We are not able to evaluate quantitatively the effect of such regions on the measurements; we think that it is small. But this means that the hydrogen masses derived from 21-centimeter data are lower limits. There is, however, an opacity effect due to the inclination of the system. Our line of sight through an edge-on galaxy is greater than through a face-on system. The situation is the same as the reddening caused by the inclination of the system. As in the optical case, this dependence on inclination can be calibrated and used to correct the data.

The late-type spirals, Sc and Sd (ordinary and barred), have 10 percent of their total mass in the form of neutral atomic hydrogen.

The irregular galaxies have about 20 percent, while the Sa's and Sb's have about 5 percent. Hydrogen-line radiation has been detected from only one elliptical galaxy, and it has much less than 1 percent hydrogen.

Other forms of hydrogen may be present: molecular (H_2) and ionized (HII). Both are known to exist in the interstellar medium of our Galaxy. Hydrogen, the most abundant element, is the principal constituent from which stars are formed. Two-thirds of the Sun's mass is hydrogen. The formation of stars and the evolution of galaxies are intimately related to the presence of hydrogen. Elliptical galaxies, devoid of those young, bright, blue stars so common in late-type spirals and irregulars, also lack hydrogen, although it is abundant in the later-type galaxies.

Does this mean that one type of system is younger than another? That one structural type evolves into another? Theories, or viewpoints at least, have been proposed suggesting a temporal relation among the different galaxy types. Different theories have different sequences of aging: in some the ellipticals are the oldest; in others the irregulars are the oldest. A third possibility is that all galaxies form at the same time, with differences of type reflecting differences in initial conditions of formation. This list is not complete; there are variants of these possibilities. In defining the problem, it is important to distinguish two types of ages: chronological and evolutionary. A dog is old at 15, although the number of years is few compared to a human life span.

Various astronomers subscribe to one or another of the above evolutionary schemes. In my opinion the present data are not sufficient to favor any particular theory. We see galaxies of different structural types—elliptical, spiral, and irregular—all of which contain stars we believe to be old. This would imply that they have the same chronological, but different evolutionary, ages.

Another important clue is that the ratio of hydrogen mass to total luminosity (or brightness) is strongly correlated with structural type and is greatest for the later systems. The ratio is a measure of the material from which stars are formed (hydrogen) to the rate at which they are slowly being "dismantled" (the luminosity). This ratio has the dimension of time, a time which tells us how long the present rate of star formation can continue.

Looking at the numbers of bright stars in a galaxy is not enough to date the system. We must consider the stars in relation to the

amount of material available for future generations of star formation. The irregular galaxies display the highest *surface density** of
young stars. But they also have the highest proportion of hydrogen
from which to continue making these stars. Unfortunately, such data
do not allow us to age-date a galaxy. Rather they suggest that
galaxies differ in the efficiency with which they convert interstellar
matter into stars, with the ellipticals being most efficient and the irregulars least efficient. An underlying, more fundamental effect presumably operates here. The hydrogen data allow us to better define
the question as to what this effect might be. Possible, but unestab-
lished, suggestions include the distributions of density and/or angular momentum.

There is an indication that the spatial distribution of the hydrogen
within a galaxy may vary with type. The data are few, because the
angular resolution available with a radio telescope is poor, and only
a few galaxies are large enough to yield good linear resolution. This
situation will change in the next few years as line-interferometric**
techniques are applied to the study of galaxies. The preliminary results indicate that the central concentration of hydrogen increases as
we go to later types: In irregular galaxies, the hydrogen appears
centrally concentrated; in earlier types this central concentration
decreases. In the Andromeda galaxy, an Sb system, there is a minimum in the hydrogen distribution in the center. A map of the hydrogen distribution gives the appearance of an annulus or ring. Because
the data are few, explanations other than type may be invoked.
Thus, the variation of hydrogen distribution can also be correlated
with the total mass of the system. The irregular galaxies have the
lowest mass and Andromeda has the greatest mass for this sample
of galaxies.

Abundance determinations for elements other than hydrogen are
available for several galaxies. The chemical composition is found
from optical studies of regions of ionized hydrogen where the atoms
of various elements are excited to high energy states and display
emission lines (*e.g.*, the Eta Carina nebula, Fig. 9). The brightness
of an emission line is related to the abundance of its atoms in a man-

---

*Surface density is number per unit area, without regard for the actual depth or
thickness of the galaxy.

**By electronically mixing the signals received by widely separated radio tele-
scopes, we may achieve angular resolutions as small as 0.0005 arc-second—1000
times better than with optical telescopes!

ner analogous to the 21-centimeter emission line. The helium abundance for eight spirals, including our own, and for a variety of types, S0 through Ir, is found to be remarkably constant, with a number ratio of helium to hydrogen atoms of 0.11. This ratio is observed to be independent of the mass of hydrogen in the galaxy, which, for different spirals, ranges from 5×10^7 to 10^{10} solar masses. The helium abundance must show a correspondingly wide range in mass content, since the helium-to-hydrogen ratio is constant.

The origin of helium in the Universe is especially interesting to cosmologists, because in any particular world model they can predict the abundance of helium. There are several mechanisms by which helium may be manufactured and put into the interstellar medium: (1) through the nuclear processes operating in the first moments of the history of a big-bang universe (see Essay 8); (2) through nuclear processes in massive objects that may have existed in the early history of galaxies; and (3) through nuclear processes in normal stars that, as they evolve, shed their helium-enriched material.

The first suggestion most easily explains the constancy of the helium-to-hydrogen ratio. The other two mechanisms require a helium production rate that is somehow coupled to the hydrogen that is already present in the galaxy. This could come about if the amount of hydrogen regulates the number of stars or massive objects that are formed, thereby determining the amount of helium manufactured. Since the hydrogen content of spiral galaxies ranges over a factor of 200, a corresponding range of 200 in helium must be manufactured in different galaxies. A rather delicate balance between the helium and the cycled and uncycled hydrogen is thus required, and theories favoring the second and third mechanisms must allow for this. Other elements, most noticeably nitrogen, do show an abundance variation from one spiral to another and especially across the face of these systems. Hence, a continual enrichment of the interstellar medium through nucleosynthesis processes in stars seems to offer a reasonable explanation.

Peculiar Extragalactic Systems

There are galaxies that defy classification in the Hubble sequence, or else have a prominent structural feature which makes their classification difficult. We use the term *peculiar galaxies* as a catch-all for these systems. The upper two photographs in Figure 22 illustrate two of the many forms of such galaxies.

It is noteworthy that most *radio galaxies*, *i.e.*, galaxies that are intrinsically strong emitters of radio radiation, fall into the peculiar category; however, the converse is not true. Many peculiar galaxies are members of a binary system or a group of galaxies, and there is a strong indication that tidal interaction is responsible for their unusual structural appearance. But tidal forces may not account for all such systems, and their peculiar aspects may hold important clues on the origin of galaxies. These galaxies may be young (chronologically), having recently condensed from the (presumed) intergalactic material. An even more exciting suggestion is that peculiar galaxies are the progenitors of other galaxies and of quasars. The latter objects are distinguished by their blue color, radio radiation, almost-starlike appearance, and high red shift.

It is the high red shift and strange nature of quasars that have caused some astronomers to question the cosmological Doppler interpretation of their red shift. If this Doppler interpretation is correct, then the quasars are very distant objects, near the observable limits of the Universe. The relation between red shift and distance is one of the cornerstones of extragalactic astronomy. It was first established in the early 1930's by Hubble and his colleague, M. Humason. But there is some disquieting evidence that the red shift of a galaxy or quasar may have a non-Doppler (and unexplained) origin. A striking example of two galaxies with a luminous bridge connecting them is shown in the lower right part of Figure 22. This system has been studied in detail by H. C. Arp of the Hale Observatories. He finds that the brighter system has a red shift corresponding to a radial velocity of 8800 kilometers per second; the fainter has a velocity of 16,900 kilometers per second. Based on these red shifts, the corresponding distances are 300 million and 550 million light-years. The two "galaxies," although apparently connected, are separated by 250 million light-years. This is over 100 times greater than the distance between the Milky Way Galaxy and the Andromeda galaxy! There are other examples of discordant velocities for apparently physically related galaxies. We do not yet understand these results, and the question of the distance scale for at least these peculiar galaxies, as well as for the quasars, must be considered open.

A rather stringent test has been applied to the form of the Doppler expression which requires that the velocity be independent of the wavelength used to measure it or, alternatively, that the amount of red shift be directly proportional to the wavelength. Red shifts for galaxies may be measured over a wavelength range of half a million: from blue light, whose wavelength is about 4×10^{-5} centimeter, to the hydrogen-line wavelength of 21 centimeters. Such a test has been made for over a hundred galaxies covering the velocity range from -400 to $+5000$ kilometers per second. For the sample, the statistical agreement is essentially perfect; the derived velocity is indeed independent of the wavelength. This does not prove that red shifts are Doppler in origin; the test is necessary but not sufficient. It does, however, require any alternative explanation of the red shift to have the same wavelength dependence, and no other satisfactory explanation has been prepared.

Most galaxies are members of physically related groups and clusters of galaxies (see the frontispiece). Our Milky Way Galaxy is a member of a smaller group of about two dozen objects; other members of this so-called Local Group are the Andromeda galaxy (Fig. 12) and the Magellanic Clouds, two small galaxies only 200,000 light-years away. Clusters are the most massive objects in the Universe, and their mass poses still another intriguing problem. The mass of a cluster may be determined by two methods. The first relates the gravitational mass that binds the system together to the observed motions of the galaxies within a cluster. The second involves estimating the mass of each cluster member from its luminosity, then summing all the masses. These two approaches should give the same answer to within the uncertainties of the data, about a factor of 2 to 5. Instead, we find differences up to factors of several hundred, in the sense that the gravitational mass is greater than the luminosity mass. There are two competing explanations. One is that the mass represented by the luminous matter—the galaxies—is only a small fraction of the total mass of a cluster. This would imply that much of the matter in the Universe is invisible to us. The alternative is to question the important assumptions made in deriving the gravitational mass of a cluster from the motions observed within it. These assumptions are (1) that the cluster is stable and (2) that the red shift, which is used to evaluate the motions, is wholly a Doppler effect.

The dilemma, often referred to as the "missing mass problem," has not been resolved. The substantiation of any of these possible explanations is of the utmost importance, but the critical test to distinguish among them has not been devised.

Another aspect of clusters returns us to the theme of structural types. A long-recognized, though poorly appreciated, property of clusters is the difference in the types of galaxies they contain. Some clusters are almost entirely populated by early-type galaxies (elliptical and S0 systems), while others have a mainly late-type membership. Clusters with a population between these two extremes are also known. An example of a mixed group of galaxy types is shown in Figure 23; one galaxy is an elliptical and the other three are spirals. Do these population differences also reflect an age, or are they an environmental effect? How can we distinguish the effects of evolution from the effects of initial conditions?

There is not yet enough information to answer these questions in any really satisfactory way. But at least we know many of the relevant physical quantities. One of the most important is the *angular momentum* of the gas from which the galaxy formed. This is, approximately, the product of the mass, the rotational speed, and the radius of the early protogalaxy. This quantity is important because, unless the protogalaxy was distorted by strong gravitational torques from nearby protogalaxies, its total angular momentum remained constant as it contracted to form the galaxy. With optical and radio telescopes, we can now measure the speed at which different parts of a galaxy rotate and estimate the total angular momentum. Spirals, not too surprisingly, tend to have more angular momentum per unit mass than do ellipticals, and could have formed in more turbulent regions of a protocluster. The rotation of galaxies is not the only clue to their origin, but it may be the most legible.

Portents of a New Astronomy

We have just mentioned where our "knowledge" is shakiest with regard to normal and peculiar galaxies. Now let us conclude by briefly looking at those extragalactic objects—exploding galaxies, radio galaxies, Seyfert galaxies, and quasars—whose properties are so strange that they almost threaten the very fabric of our science.

Figure 15 shows the peculiar galaxy Messier 82 in hydrogen light. We see vast masses of hydrogen gas rushing away from the galactic nucleus, where an explosive event of extraordinary violence appears to have taken place within the last million years. Exploding galaxies, such as this one, are not easily fit within our preconceived conservative notions of galaxies as long-lived, slowly evolving entities in the Universe.

Our initial concepts must be amended to include previously unsuspected violence in galactic nuclei; phenomena involving exceptionally high energy production; and the possibility of drastic evolutionary effects in galaxies. *Radio galaxies* were, perhaps, the first clues to suggest that our ideas needed revision. These powerful radio sources are apparently galaxies that have ejected large clouds of high-energy elementary particles, which emit intense synchrotron radiation (see Essay 3) as they spiral along lines of magnetic force. About a dozen so-called *Seyfert galaxies* have been found. These are spirals distinguished by a bright starlike nucleus; violent gas motions, high temperatures, and short-term variability characterize these galactic nuclei.

Finally, there are the stranger-than-fiction *quasars*, whose very high red shifts we have already mentioned. These objects are seen almost to the "edge" of the known Universe, where they are receding from us at about the speed of light (if we can believe the Doppler shift interpretation of their spectral-line displacements). Though they appear starlike, they must be shining with the energy of 10,000 galaxies to be as bright as they appear at their great distances! But they do not shine steadily, for dramatic light variations are evident within months; therefore, the characteristic size of their emitting regions must be less than one light-year across. Radio-interferometric investigations have revealed (1) that the extraordinary energy of a quasar does emanate from a region smaller than a light-year across, (2) that even the radio emission is grossly time-variable, and (3) that some quasars have cores that seem to be expanding at apparent speeds greater than the speed of light (see Fig. 16). These objects are extremely bright at ultraviolet and infrared wavelengths, and they exhibit absorption red shifts that often differ from their emission red shifts.

Many theories have been advanced to explain the quasars, but none has yet met with universal acceptance. Some have suggested that part of their red shift is intrinsic, *i.e.*, not caused by the Hubble expansion, so that these objects are rather nearer, and less powerful; but no acceptable theory of intrinsic red shifts has been presented. The conservative viewpoint is that they are, indeed, distant and powerful; condensed massive objects that represent a short (million years?) active phase of the evolution of some galactic nuclei. Among the mechanisms proposed for explaining their great energy output are: (1) superdense star clusters with many stellar collisions or supernovae; (2) a supermassive evolving "star," rotating disk, or magnetized plasma cloud—all around 10^6 to 10^8 solar masses; (3) a dense cluster of pulsars, *i.e.*, rotating neutron stars; (4) the initial burst of star formation in a newly formed galactic core; and (5) the late emergence of part of the Universe as a "white hole" (the time-reverse of a collapsing "black hole") of galactic mass.

Speculation is free, but understanding requires hard work and solid data. Only by careful and prolonged observation will we resolve the enigma of the quasars and finally be able to say that we understand the "landmarks" of our Universe.

This survey has covered a period of approximately 2500 years, from the suggestions of Democritus on the composition of the Milky Way to some of the most recent research on galaxies and strange extragalactic objects. Of this time span, the extragalactic era is only about 50 years old. I have touched upon only a few of the many aspects of research in this era, the choice often representing my own views and interests. I have purposefully concluded with a variety of topics which emphasize our lack of knowledge. We hope that the data at least allow us to ask the right questions, the answers to which will make the next 50 years even more exciting than those just past.

7. What Olbers Might Have Said

Peter T. Landsberg and David A. Evans
University College, Cardiff, Wales

1. *Introduction*

After the invention of the telescope, it became evident that the number of stars in existence was very large and that, assuming their intrinsic brightnesses were about the same as the Sun's, they must be distributed more or less at random through a very large volume of space to account for the differences in observed brightness. The question naturally arose whether the total number of stars was finite or infinite. A possible answer seemed to come from Isaac Newton's law of gravitation. If a finite number of stars are placed at rest in an otherwise empty, infinite Euclidean space, they will attract each other gravitationally and will therefore ultimately collide. Philosophical problems also arise; if the Universe is finite and surrounded by infinite space, why should it be in one part of that space rather than another, and is it meaningful to ask this question? The alternative picture, of an infinite universe uniformly filled with stars, was therefore more attractive and was widely held to be correct in the eighteenth century. However, P. L. de Cheseaux (in 1744) and H. Olbers (in 1823) showed that the assumption of an infinite universe could also lead to a paradoxical conclusion, namely, that the sky ought not to be dark at night! In the following sections we shall restate the paradox and some possible resolutions of it. In particular, we shall ask by what routes Olbers, using the physical theories of his time, could have arrived at the present-day view that the sky is dark at night because the Universe is expanding. This question represents one of the earliest quantitative attempts to understand cosmology. As a result, our essay will provide an interesting contrast with the more qualitative descriptions of modern cosmology presented in the following essays.

D. Sciama (1959) has remarked that "Olbers could have predicted the expansion of the universe, and even made a rough estimate of

Hubble's constant, a hundred years ahead of the observers. His failure to do so is one of the greatest missed opportunities in the whole history of science." This interesting statement may be considered to have prompted our essay. It raises two questions: (1) Could Olbers have predicted the expansion of the Universe as a means of avoiding his paradox? The answer we find is that if he had postulated a Hubble-type expansion, a wide choice of additional assumptions would have been open to him. In section 5 we show that these assumptions would quite possibly have led to a finite rate I_{exp} of energy receipt per unit area from the night sky, thus resolving the paradox. However, we also show that there are some plausible models that would have left the paradox intact, and so might have led Olbers to reject the expansion hypothesis.

The second question is: (2) Could Olbers have estimated the Hubble constant H? On this point we are more sceptical. We shall establish the relevant expression in equation (5.1), which shows that this calculation would require an experimental value for the light energy, I, received per unit area per second from the night sky (appropriately corrected for nearby light sources, atmospheric absorption, and so on). It also requires a knowledge of L, the mean light output of galaxies, and also of N, their mean number density in space. Since the values of I, L, and N were uncertain by many orders of magnitude before the present century (Essay 6 has pointed out that it was not even known whether the galaxies were distant star systems or clouds of gas within the Milky Way), any calculated value of H would have been equally uncertain.

2. *The Paradox*

A Newtonian universe will be postulated. In addition certain assumptions will be introduced, each of which is, strictly speaking, regarded as in error in present-day physics. Nonetheless, at least assumptions (i) to (iii) are often postulated, even now, in the belief that, for certain broad considerations, the resulting model will be sufficiently close to reality. The approximations are:

(i) Stars are distributed uniformly throughout infinite Euclidean space and are (on average) identical to each other and to the Sun.

(ii) Each star has been shining for an infinite past time with its present brightness.

(iii) There is no light-absorbing material (dust and so on) between the stars.

(iv) There are no large-scale motions of the stars.

One can now argue as follows (for a more detailed popular exposition see Sciama [1959]). First, according to classical optics, the apparent brightness per unit area of a light source of finite area does not depend on its distance, in the absence of absorbing material. For instance, if the Sun were removed to twice its present distance the light reaching an observer's eye from the Sun would be reduced to one quarter of its present intensity, but the area covered by the Sun's image on the observer's retina (or on a photographic plate) would also be reduced by the same factor, so that the ratio of light received to area of image would be unchanged. It follows from assumption (i) that the apparent brightness per unit area of the surface of every star, at whatever distance, should be the same as that of the Sun.

Second, any straight line drawn through an infinite space containing a uniform density of finite stars will sooner or later intersect the surface of at least one star. This is the converse of the familiar observation that a sufficient depth of fog (which consists effectively of spherical drops distributed uniformly in air) will absorb and scatter all light aimed in a given direction. It follows that a line of sight, drawn in any direction from any point in the universe, will sooner or later end on the surface of a star. The apparent brightness of the sky in that direction will therefore in virtue of assumption (i) be the same as that of the Sun's surface. Since this argument applies to any direction, we conclude that the whole of the sky should appear as bright as the Sun's surface.

If this argument is accepted, at least one of the assumptions (i) to (iv) above must be discarded. We shall consider them in turn.

One way of discarding assumption (i) is, of course, to return to the finite universe considered briefly in the first paragraph. A more subtle way was pointed out by C. V. L. Charlier (1922). If we imagine that stars are grouped in clusters, that the clusters are themselves grouped in superclusters, the superclusters in higher-order clusters, and so on, it is also possible to imagine that while the local density of stars, as observed in our telescopes, corresponds to the density within a cluster and is nonzero, the large-scale density is zero. More accurately, the average density in a volume of space, V, tends to zero as V becomes infinitely large. In such a universe the chance that a line of sight chosen at random will end on a star can be made small enough to agree with observation. Recently G. de Vaucouleurs (1970) has presented observational evidence for high-order clustering, but there is no consensus of expert opinion yet on this matter (W. H. McCrea 1970).

In terms of present-day physics, assumption (ii) is known to be false (as was pointed out by E. R. Harrison in 1965) since it violates the principle of conservation of energy. Nonetheless, it represents a reasonable simplification; furthermore, it is necessary to the paradox because of the time taken for light to travel. If, to take an extreme case, all the stars in the Universe were created N years ago out of nothing, then no star at a distance from a given observer, O (who could, for example, be on the Earth), greater than N light-years distant could be seen by O at the present day. As far as visual observations are concerned, the Universe would be a finite sphere centered on O with a radius of N light-years. Very large values of N, much larger than current estimates of the age of the Sun, would be needed before the calculated brightness of the night sky became embarrassingly large.

Olbers thought that the paradox could be resolved by discarding assumption (iii), *i.e.*, by postulating large amounts of light-absorbing dust between the stars. Given the other assumptions, however, the paradox remains, because in an infinite past time the dust would have had to absorb an infinite amount of radiant energy and therefore would be infinitely hot. In practice an equilibrium state would be reached in which the dust reradiated as much energy as it absorbed, and the brightness of the sky would be unaltered.

The necessity for assumption (iv) is less obvious than for assumptions (i) and (ii). It arises from the fact that, for a number of reasons, the energy density at a receiver of light received from a receding source is smaller than it would be if the same source were at rest. This effect is particularly strong if the distant stars are receding from the Earth with speeds comparable to that of light, even in the framework of prerelativistic physics, and it would reduce the expected brightness of the sky. This remark will be made more precise in section 4. Before doing so, it will be useful to restate the paradox in a form more suited to discussing the various effects of the recession.

Consider a shell bounded by two spheres centered on the observer O (see Diagram 7.1). The radii of the spheres are r and $r + h$. We assume h to be much smaller than r, so that

(i) area of outer sphere $\simeq$ area of inner sphere = $4\pi r^2$, and so

(ii) volume of shell $\simeq h \times$ area of spheres $\simeq 4\pi r^2 h$.

Hence, if N is the uniform number density of stars in space, then the

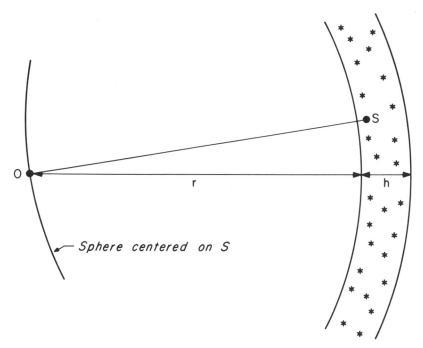

Diagram 7.1. A thin spherical shell of stars (or galaxies) is shown in part, centered upon the observer *O*.

number of stars in a shell of inner radius *r* is

$$4\pi N r^2 h \equiv n(r). \tag{2.1}$$

Now consider how much light is received from any one of these stars by a collecting surface of unit area near *O*. If we draw a sphere centered on the star *S* and passing through *O*, the total light output of the star (say *L* ergs per second) must be spread uniformly over the sphere, whose surface area is $4\pi(OS)^2 \cong 4\pi r^2$. The collecting surface of unit area near *O* represents only a fraction, $1/(4\pi r^2)$, of the sphere, and so receives only this fraction of the light output. Hence, the light received near *O* from a single star in the shell is

$$I = \frac{L}{4\pi r^2} \text{ ergs cm}^{-2} \text{ sec}^{-1}. \tag{2.2}$$

If all the $n(r)$ stars in the shell have the same light output *L*, then evidently the light received near *O* from all stars in the shell is

$$I(r) = n(r)\frac{L}{4\pi r^2} \text{ ergs cm}^{-2} \text{ sec}^{-1}. \tag{2.3}$$

In practice L must be an average light output for stars in the shell. Notice that we have added together in equation (2.3) light coming to O from all directions. This corresponds to a collecting surface which is a sphere of unit cross section.

The paradox now follows if we note that the combination of equations (2.1) and (2.3) yields

$$I(r) = LNh. \tag{2.4}$$

Thus each of an infinite number of shells of thickness h contributes the same finite amount of light LNh. The total amount of light received is, assuming N to be independent of r,

$$I_{total} = I(O) + I(h) + I(2h) + \cdots$$
$$= LNh + LNh + LNh + \cdots = \infty. \tag{2.5}$$

This simple calculation differs from the previous nonmathematical argument in that it does not allow for the absorption of light from distant stars by nearer stars. The effect of this absorption is to reduce the predicted brightness of the sky from infinity to the brightness of the Sun's surface. Note also that the first few terms in the sum (2.5) are inaccurate, since equation (2.1) is based on the assumption that h is much smaller than r, which is evidently false for $r = O, h, 2h$. This inaccuracy is unimportant because most of the light contributing to I_{total} comes from very large distances, and the difficulty can in any case easily be overcome by use of the calculus.

Any resolution of the paradox must modify equation (2.4) so as to reduce $I(r)$ for large r. This modification can be expressed by

$$I(r) = L N'_r f_r h, \tag{2.6}$$

where N'_r denotes the true density of stars at distance r. The adjective "true" is necessary because one effect of the expansion of the Universe is that it leads to an apparent density which differs from the true density (as seen in equation (4.20)). The factor f_r includes all the effects on $I(r)$ of the expansion. For a finite static universe of radius R and center O, equation (2.6) still holds with

$$f_r = \begin{cases} 1(r \le R) \\ O(r > R) \end{cases}. \tag{2.7}$$

A finite, uniform, and static universe is therefore included among the possibilities if we use equation (2.6) with equation (2.7) and set N'_r equal to N, a constant. In that case, the total energy received at O per square centimeter per second is

$$I_{\text{static}} = LNR. \tag{2.8}$$

This must be compared with an expanding universe, for which one has from equation (2.6)

$$I_{\text{expand}} = L \int_0^\infty f_r N_r' dr. \tag{2.9}$$

The radius of an equivalent static universe is therefore seen to be

$$R_{\text{equiv.}} = \frac{1}{N} \int_0^\infty f_r N_r' dr. \tag{2.10}$$

Although equations (2.7) and (2.8) apply to a static universe of finite extent but infinite age, a very similar result holds for a static universe of infinite extent but of finite age T. In that case light traveling at the speed c from a source at a distance greater than cT from the observer has not yet had time to reach him. Only sources at distances less than or equal to cT are visible to the observer, and this distance therefore presents a kind of horizon. One finds in analogy with equation (2.8),

$$I_{\text{static, finite age}} = LNcT. \tag{2.11}$$

In such a universe the background brightness of the sky would increase steadily with time T until, in a very distant future, it became equal to the brightness of the Sun's surface.

3. The Simpler Models Open to Olbers

In this section we enumerate the simpler theories that Olbers could have considered. These theories have two aspects. First, it is necessary to choose a cosmological model to describe the overall structure and evolution of the Universe. Second, it is necessary to choose a model for the propagation of light in the Universe. The combination of these two aspects then yields a simple theory in terms of which calculations can be made.

a. Simplest Cosmological Models

Olbers considered a universe in which the number density of stars was uniform. The simplest model to take account of an expansion is that which retains the uniform density of stars and adds a velocity of recession proportional to distance:

$$N'_r = N, v_r = Hr, \qquad (3.1)$$

where N'_r is the density of stars in shell r, v_r is the speed of recession of these stars, and N and H are constants.

Since the observer sees distant shells as they were in the past because of the time taken for light to travel, the assumption $N'_r = N$ means that the density of stars has not varied with time in the past. Assumptions (3.1) thus represent a steady-state universe, as advocated by F. Hoyle (1948). The alternative assumption of a big-bang universe, in which the density was very high at a definite time in the past, evidently replaces equations (3.1) with more complicated expressions. We shall confine our arguments to equations (3.1), although the big-bang model might well have appeared just as plausible to Olbers.

b. *Simplest Models of Light Propagation*

There are three possible models of light propagation:

I. Corpuscular theory: The simplest model will be adopted here, namely, that the speed of the light corpuscles relative to the star is a constant, to be denoted by c. In addition it is convenient to assume that the light emission is isotropic in a reference frame in which the star is at rest.

II. Wave theories with an ether which does not participate in the expansion of the universe, so that stars move through the ether. In this case the speed of the light is a constant relative to the ether, to be denoted by w_{LE}.

    A. Emission is isotropic with respect to the ether; the motion of the receiver is arbitrary.

    B. Emission is isotropic with respect to the ether; the receiver is assumed at rest in the ether. This is a special case of A.

    C. Emission is isotropic with respect to the source; the motion of the receiver is arbitrary.

    D. Emission is isotropic with respect to the source; the receiver is at rest in the ether. This is a special case of C.

III. Wave theories with an ether which expands with the universe, so that the stars are at rest in the ether.

The third group of theories will not be considered in detail, for it had always been thought that the ether did not participate in the motion of solid bodies such as the Earth (the phenomenon of the aberration of starlight was only explainable by this view). We shall however note that these theories result, with equations (3.1), in a finite radius, $R = c/H$, for the observable universe, since the light emitted by a source at distance $r > R$ experiences an "ether headwind" of speed $v_r = Hr > c$ which carries it away from the observer at a net speed $(v_r - c) = H(r - R)$. It follows that the number of sources from which light can reach the observer is finite, and therefore so is the total amount of light received by him. The choice of a theory of light propagation of type III therefore resolves the paradox.

It should be noted that the particle (*i.e.*, corpuscular) theory of light had had support, for example, from Maupertuis's explanation of the refraction of the ordinary ray in crystals (1744) and from Laplace's analogous explanation of the extraordinary ray (1809). Young's explanation of Newton's rings (1801) and Fresnel's explanation of diffraction (1816, 1826), on the other hand, had supported the wave theory. A historical account is given by E. T. Whittaker (1951).

The steady-state model, attributed here hypothetically to Olbers, implies a generation rate of galaxies which operates to keep the density of galaxies constant, although the volume of space bounded by any set of existing galaxies is continually increasing with time.

This generation rate is most easily calculated by reference to Diagram 7.1, which must now be interpreted as showing two stages in the history of an expanding universe (we must also replace the word *star* with *galaxy*). Consider a set $[S]$ of galaxies which at time t lie on a sphere of radius r centered on O. At a later time, $t + \Delta t$, each galaxy will have receded a distance which, if Δt is small, can be expressed as $v_r \Delta t$ or, by equations (3.1), as $Hr\Delta t$. If we identify h of Diagram 7.1 with $Hr\Delta t$, we can say that each of these galaxies, which at time t was on the inner surface of the shell illustrated, has receded a distance h at time $t + \Delta t$ and is therefore on the outer surface of the shell. Thus, the volume of space bounded by the set $[S]$ of galaxies at time t is

$$V(t) = \frac{4}{3} \pi r^3, \tag{3.2}$$

and the volume of space bounded by $[S]$ at time $(t + \Delta t)$ is

$$V(t + \Delta t) = V(t) + V(\text{shell}) \cong V(t) + 4\pi r^2 h$$
$$= V(t) + 4\pi r^3 H \Delta t \qquad (3.3)$$
$$= V(t) + \Delta V.$$

But according to assumption (3.1) the number density of galaxies, N, is constant in space and time. Thus, if the volume bounded by the galaxies [S] has increased by ΔV, the number of galaxies in this volume must have increased by $N\Delta V$. Since no galaxies enter this volume from outside, the increased number must be due to a generation rate, g, of galaxies throughout the volume. Formally,

$$g = \frac{N\Delta V}{V(t)\Delta t} = \frac{N 4\pi r^3 H \Delta t}{\frac{4}{3}\pi r^3 \Delta t} = 3NH. \qquad (3.4)$$

Taking the numerical estimates $N \sim 3 \times 10^{-74}$ galaxies per cubic centimeter, $H \sim 1/3 \times 10^{-17}$ per second, and $M \sim 10^{45}$ grams as a typical mass of a galaxy, we find

$$g \sim 3 \times 10^{-91} \text{ galaxies cm}^{-3} \text{ sec}^{-1}$$
$$\sim 5 \times 10^{-46} \text{ g cm}^{-3} \text{ sec}^{-1},$$

which equals approximately one hydrogen atom in a cube of side 10 meters in a century.

On the corpuscular theory of light, galaxies receding with a speed in excess of the light speed c would be invisible since the light from them would never reach O. The radius of the observable universe is then

$$\frac{c}{H} \sim 10^{28} \text{ cm} \sim 10^{10} \text{ light-years.}$$

The mass contained in it would be

$$\frac{4\pi}{3} NM \left(\frac{c}{H}\right)^3 \sim 10^{56} \text{ g} \sim 5 \times 10^{22} \text{ solar masses,}$$

and this figure remains constant in time because of the assumed exact balance between the creation rate and the loss of galaxies beyond the range of observation.

4. *Factors Affecting the Amount of Light*
Received from Receding Sources

It is possible to distinguish three different effects of recession upon light:

(i) the *Doppler effect*, which alters the apparent frequency (*i.e.*, distribution in time) of the light received from each source,

(ii) the *aberration effect*, which alters the apparent distribution in space of the light received from each source,

(iii) the *density transformation*, which alters the apparent density in space of these sources.

If we assign factors f_D, f_a, and f_d to these effects, then equation (2.3) is to be replaced by

$$I(r) = \frac{n(r)}{4\pi r^2} L f_D f_a \text{ ergs cm}^{-2} \text{ sec}^{-1} \qquad (4.1)$$

and equation (2.1) by

$$n(r) = 4\pi f_d N'_r r^2 h, \qquad (4.2)$$

equation (4.1) signifying that the light received from each star is reduced by a factor $f_D f_a$, and (4.2) that the apparent density of stars is $N_r = f_d N'_r$. Combining equations (4.1) and (4.2) gives an equation of the form (2.6):

$$I(r) = f_D f_a f_d L N'_r h = f_r L N'_r h. \qquad (4.3)$$

We shall now identify the factors f_D, f_a, and f_d for the simplest possible model of light propagation, that of the corpuscular theory (model I). Results for other models will be discussed briefly where appropriate.

(i) *The Doppler Effect.* Suppose that a star S is receding from the receiver O with constant speed v. At time $t = 0$, S emits a corpuscle which reaches O at time t_1. At time $\Delta't$, S emits a second corpuscle which reaches O at time t_2. Assume that the distance of S from O is r when the first corpuscle is emitted. This is shown in Diagram 7.2 (drawn from O's viewpoint). We now argue as follows. Let w be the speed of the corpuscles relative to O. Then the time taken for the corpuscles to travel from S to O is

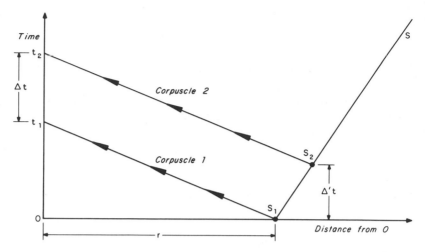

Diagram 7.2. The classical Doppler effect arises from the time delay shown here.

$$t_1 = \frac{OS_1}{w} = \frac{r}{w}$$

$$t_2 = \Delta't + \frac{OS_2}{w}.$$

(4.4)

But the star, moving at speed v relative to O, covers the distance S_1S_2 in time $\Delta't$, so that

$$OS_2 = OS_1 + S_1S_2 = r + v\Delta't.$$

(4.5)

Substituting this in equations (4.4) gives

$$t_1 = \frac{r}{w}$$

$$t_2 = \Delta't + \frac{r + v\Delta't}{w} = \frac{r}{w} + \Delta't\left(1 + \frac{v}{w}\right).$$

(4.6)

But the time interval between emission of corpuscles is $\Delta't$, while the time interval between reception of the same corpuscles is $\Delta t = t_2 - t_1$. From equations (4.6) we find,

$$\Delta t = t_2 - t_1 = \Delta't\left(1 + \frac{v}{w}\right).$$

(4.7)

Now if corpuscles 1 and 2 are successive corpuscles emitted in the required direction, then the frequency of emission of corpuscles is $\nu' = 1/\Delta' t$ corpuscles per second and the frequency of reception of corpuscles is $\nu = 1/\Delta t$ corpuscles per second, and so, using equation (4.7),

$$\frac{\nu}{\nu'} = \frac{\Delta' t}{\Delta t} = \frac{1}{\left(1 + \dfrac{\nu}{w}\right)}. \tag{4.8}$$

This result may be developed further by noting that, in the emission or particle theory, all particles are emitted with the same speed, to be denoted by c, relative to S. The velocity of a particle is along the line SO, and S has a speed v in the opposite direction relative to O. Therefore,

$$w = c - v, \tag{4.9}$$

so that

$$\left(\frac{\nu}{\nu'}\right)_{\mathrm{I}} = \frac{1}{1 + \dfrac{v}{c - v}} = \frac{c - v}{c - v + v} = 1 - \frac{v}{c}. \tag{4.10}$$

A particularly simple qualitative explanation of the Doppler effect can be given for the corpuscular theory: each corpuscle has to cover a slightly greater distance than its predecessor in order to reach O and is therefore delayed in transit by a slightly longer time, so that the time interval between arrivals is greater than that between departures.

It remains to identify the factor f_D. For the corpuscular theory (model I), each corpuscle carries a definite energy E, so that the rate of emission or reception of energy is simply proportional to the numbers of corpuscles emitted or received in unit time, *i.e.*, to the frequencies. Hence,

$$f_D(\mathrm{I}) = \left(\frac{\nu}{\nu'}\right)_{\mathrm{I}} = 1 - \frac{v}{c}. \tag{4.11}$$

An alternative assumption would be that each light corpuscle (of mass m) is brought to rest relative to O and gives up its kinetic energy $1/2\, mw^2$. This energy is smaller than would be the case for a source at rest by the factor w^2/c^2, leading with equation (4.9) to $f_D(\mathrm{I}) = (1 - v/c)^3$. This assumption would strengthen our conclusion that model I leads to a finite brightness of the night sky.

Turning to the wave theories, observe that the argument up to equation (4.8) holds for model II if one replaces "corpuscle" by, say, "wave crest in the ether." The speed w is now that of the light waves relative to O. Fresnel had explained the refraction of light by the wave theory (1823), and the notion of waves of intensities a^2 and b^2 leading to maximum and minimum intensities $(a \pm b)^2$ was current in Olbers's lifetime. He might therefore possibly have argued, had he used model II, as follows: the mechanical energy of a vibrating system is proportional (other things being equal) to the square of the frequency, and the same is true of light considered as a vibration in the ether. Hence

$$f_D(\text{II}) = \left(\frac{\nu}{\nu'}\right)^2_{\text{II}} = \frac{1}{\left(1 + \dfrac{\nu}{w}\right)^2}. \qquad (4.12)$$

The argument leading to equation (4.8), and certainly equation (4.9), would have been just within the reach of astronomers or physicists at the time of Olbers. Indeed, as a matter of historical fact, the Doppler effect for sound and light was predicted by Doppler in 1842, only two years after Olbers's death. The experimental confirmation by Buys Ballot for sound came in 1845. We shall return to the question whether the discovery of the expansion of the Universe by Olbers on the basis of his paradox would have required necessarily the incidental discovery of the Doppler effect.

(ii) *The Aberration Effect*. One is historically on a sound basis in assuming that Olbers would have introduced corrections due to the aberration of light in any amended argument that allowed for an expansion of the Universe. Bradley had discovered aberration in 1727, and its implications were of importance to astronomers.

A star S moving radially with respect to the receiver O may emit radiation at an angle θ' with respect to the line OS. Furthermore, let the speed of the light signal with respect to the star be w'. Then these quantities with respect to O may be denoted by θ and, as in the section on the Doppler effect, by w. It is clear from velocity addition that the angle θ is greater than θ'. This means that a given amount of energy radiated in one second will be distributed over a greater angle when the process is viewed at O than when it is viewed at S. Thus, although the emission is isotropic for anyone on S (by assumption in the model I), it is no longer so for O: he sees a more dilute energy density in the wake of the source.

The effect is calculated by observing from velocity addition (Diagram 7.3) for the corpuscular theory that

$$w' \cos \theta' = w \cos \theta + v$$
$$w' \sin \theta' = w \sin \theta. \tag{4.13}$$

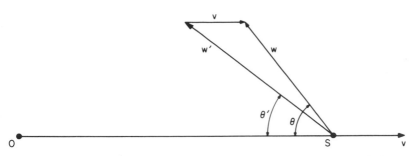

Diagram 7.3. The aberration of velocities arises when the source S moves relative to the observer O, while emitting particles or radiation at some angle to the line of motion.

In fact only small angles θ are of interest if the emitted light is to hit a surface at O a long distance from S. Hence

$$\frac{\sin \theta'}{\sin \theta} = \frac{w}{w'} = \frac{\cos \theta'}{\cos \theta + \dfrac{v}{w}} \tag{4.14}$$

becomes

$$\frac{\theta'}{\theta} \simeq \frac{1}{1 + \dfrac{v}{w}}. \tag{4.15}$$

Consider now a given conical pencil of radiation of angle θ' as judged by S and emitted by S under two conditions: (a) in the absence of expansion, and (b) when S recedes from O, at the instant when the distance OS is r. This is illustrated in Diagram 7.4 from O's viewpoint. The (circular) areas hit by the pencil at O are in the two cases

$$\pi a'^2 = \pi r^2 \theta'^2 \text{ and } \pi a^2 = \pi r^2 \theta^2$$

respectively. The area is larger in the case of expansion, and this implies an energy dilution factor at O due to aberration. The energy density received at O in the presence of expansion is therefore less than that in the absence of expansion by a factor

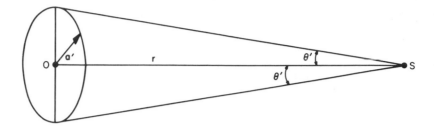

O's View

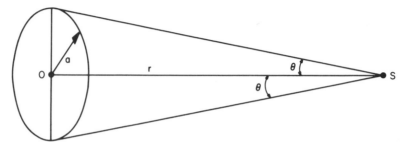

Diagram 7.4. The aberration dilution of radiant energy in the wake of a receding source is a consequence of the aberration of velocities.

$$f_a(\mathrm{I}) \equiv \frac{\pi a'^2}{\pi a^2} = \left(\frac{\theta'}{\theta}\right)^2 = \frac{1}{\left(1 + \dfrac{v}{w}\right)^2}. \tag{4.16}$$

One can again use equation (4.9) for the case of model I of section 3, since expressions (4.13) hold only if w and v are calculated with reference to the same system, and this is the receiver O in this case. Hence

$$f_a(\mathrm{I}) = \left(1 - \frac{v}{c}\right)^2. \tag{4.17}$$

Turning to the wave theories (model II), it is clear that an ether wind will in general affect the velocity of the signals relative to O and, furthermore, the importance of this effect will depend on which stars have emitted the signal, since the different stars have different speeds in the ether in virtue of the cosmological assumption (3.1). The problem is tractable in terms of the velocity of the light in the ether (w_{LE}), the velocity of O in the ether (v_{OE}), and the velocity of the star relative to O (v_{SO}), and yields for $v_{OE} < w_{LE}$

$$f_a(\text{IIA, IIB}) = \frac{w_{LE}(w_{LE}^2 - \mathbf{v}_{OE} \cdot \mathbf{w}_{LE} - \mathbf{v}_{SO} \cdot \mathbf{w}_{LE})}{|\mathbf{w}_{LE} - \mathbf{v}_{OE} - \mathbf{v}_{SO}|^3}. \quad (4.18)$$

If O is at rest in the ether, this equation simplifies since $\mathbf{w}_{LE} = \mathbf{w}_{LO}$ and $|w_{LO}|$ has already been given as the interpretation of w in the wave theories. Also $\mathbf{v}_{SO}$ and $\mathbf{w}_{LE}$ are in opposite directions, so that $\mathbf{v}_{SO} \cdot \mathbf{w}_{LE} = -vw$. Thus one obtains a formula identical with equation (4.16):

$$f_a(\text{IIB}) = \frac{1}{\left(1 + \dfrac{v_{SO}}{w_{LO}}\right)^2}. \quad (4.19)$$

In both cases the speed of the light relative to O is denoted by w. In equation (4.16) this is the speed of the corpuscles. In the special case of equation (4.19), w_{LO} is the speed of the waves in the ether and, therefore, their speed relative to O.

(iii) *The Density Transformation.* This transformation has not been mentioned by any of the authors who have written about Olbers's paradox so far as we know. Its origin is best explained by imagining a rod AB receding from an observer O, in a direction along its length and at speed v. A photograph taken by O at time t^* contains information from all points of the rod, including the end points A and B. But in model I, the light signals arriving at O at time t^* arrive simultaneously and were therefore emitted at different times. This leads to a distorted view of the rod in the photograph. Since the more distant end point, B say, of the rod is seen at an earlier time (t_1) in its recessional motion (when it was closer to O) than the nearer end point A, which is seen at time t_2, the distortion in the photograph is in fact a contraction.

To estimate this effect quantitatively, let A_1B_1, A_2B_2 be the positions of the rod at times t_1 and t_2, respectively, and let the representation of the paths of A, B, and the light particles be as shown in Diagram 7.5. On this diagram the two angles α and β satisfy

$$\tan \alpha = \frac{1}{v}, \ \tan \beta = \frac{1}{w}.$$

There are also three unknown quantities, f, g, and h. The photographed length, L, is clearly given by g, and the real length of the rod is

$$L' = f + g = f + L.$$

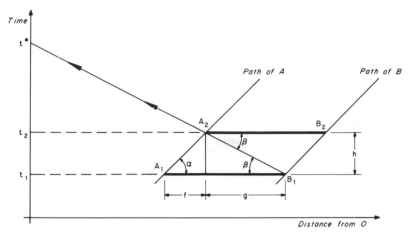

Diagram 7.5. The apparent contraction of a receding rod, moving parallel to its length, is an effect of the finite light travel time from the front to the back of the rod.

Now

$$f = \frac{h}{\tan \alpha} = g \frac{\tan \beta}{\tan \alpha} = L \frac{v}{w}.$$

It follows that

$$\frac{L'}{L} = 1 + \frac{f}{L} = 1 + \frac{v}{w}.$$

In three-dimensional recessional motion the two directions at right angles to the given direction OS do not contribute to f_d, and our final result (see equation (4.20)) remains valid.

This seen contraction leads to shorter distances between stars on photographs, and hence to higher densities in accordance with

$$f_d(\mathrm{I}) = \frac{N_r}{N'_r} = \frac{L'}{L} = 1 + \frac{v}{w} = \frac{1}{1 - \dfrac{v}{c}}. \qquad (4.20)$$

In the wave theories the analogous result, in the notation (4.18), is

$$f_d(\mathrm{II}) = 1 - \frac{(\mathbf{v}_{SO} + \mathbf{v}_{OE}) \cdot \mathbf{w}_{LE}}{w_{LE}^2}. \qquad (4.21)$$

In the special case $\mathbf{v}_{OE} = 0$, $\mathbf{w}_{LE} = \mathbf{w}_{LO}$, and $\mathbf{v}_{SO}$ and $\mathbf{w}_{LO}$ are anti-parallel. Hence we find in analogy with equation (4.19) that

$$f_d(\text{IIB, IID}) = 1 + \frac{v_{SO}}{w_{LO}}. \tag{4.22}$$

5. Models Which Olbers Might Have Used

For the corpuscular theory of light one can now use equations (4.11), (4.17), and (4.20) in equation (4.3) to find the rate of energy received at O per unit area, if N is a constant. One finds that

$$I(r)_1 = \left(1 - \frac{v}{c}\right)\left(1 - \frac{v}{c}\right)^2 \frac{1}{\left(1 - \frac{v}{c}\right)} LNh = \left(1 - \frac{v}{c}\right)^2 LNh$$

for the energy from all stars in a distance range $(r, r + h)$ from O. We now make the cosmological assumption (3.1) and sum all contributions from the various shells. For this purpose observe that by equation (4.9) no signals will be received at O from stars whose distance satisfies $Hr > c$.

As observed earlier, Olbers, or a scientist of his time, might not have considered the correction arising from the Doppler effect. Suppose, therefore, that his resulting integral had the form of equation (2.9)

$$I_{\exp}(\text{I}) = LN \int_0^{c/H} \left(1 - \frac{H}{c} r\right)^s dr = \frac{1}{s + 1} \frac{c}{H} LN. \tag{5.1}$$

It will be seen that this is always finite, for $s > -1$, and corresponds to an equivalent static and finite universe with a radius, measured from O as center, given by equation (2.10) as

$$R_{\text{equiv}}(\text{I}) = \frac{c}{s + 1} H. \tag{5.2}$$

Thus if Olbers had made the cosmological assumption (3.1), had adopted the corpuscular theory, and had corrected only for aberration ($s = 2$), he would have avoided the paradox within the terms of the physics then available.

One may go further and observe that even if he had made no correction at all ($s = 0$) except to note that on the Hubble law assumption there is in effect a finite extent $R = c/H$ to the observable Universe, he would have been able to avoid his paradox. If, however, he had introduced only the density correction factor f_d (leading to

$s = -1$ in equation (5.1)), then the integral would be infinite and the paradox would remain. This is unlikely, since the aberration effect was so well known in other contexts.

The wave theory leads to more complicated equations and integrations. We will not give the details here, but note that for emission that is isotropic with respect to the emitting star (model IIC), one has, using equations (4.12), (4.18), and (4.21) for $v_{OE} < w_{LE}$,

$$R_{\text{equiv.}}(\text{IIC}) = \frac{w_{LE}}{2H} \left[1 - \frac{2}{3} \left(\frac{v_{OE}}{w_{LE}} \right)^2 \right]. \qquad (5.3)$$

If some imperfection in Olbers's hypothetical argument made it necessary to replace the second power in equation (4.12) with the power s, one finds

$$R_{\text{equiv.}}(\text{IIC}) = \frac{w_{LE}}{sH} \left[1 - \frac{2}{3} \left(\frac{v_{OE}}{w_{LE}} \right)^2 \right]. \qquad (5.4)$$

Thus the total rate of energy received per unit area at O is still finite, provided only $s > 0$, though the range of integration leading to equations (5.3) and (5.4) is infinite.

There is one plausible class of models of a Newtonian expanding universe that Olbers might have tried without resolving his paradox. These are the models in which light propagation is isotropic in the ether (models IIA and IIB). For these models, the dilution of light in the wake of a receding *source*, described in equation (4.15), does not occur. There is, however, still in model IIA an aberration factor f_a resulting from the *observer's* motion in the ether. This factor does not, as in equations (4.17) and (4.18), become small for the more distant (and therefore more rapidly moving) stars. Although it depends only on the observer's speed in the ether, it affects light from different directions differently and in fact enhances the energy received from stars toward which the observer is moving (the so-called headlight effect). The result is that the energy density I_{exp} (IIA, IIB) predicted by these models, and hence the predicted brightness of the sky, is infinite. Olbers might have tried these models from an analogy with sound propagation.

We should also note that an additional model is possible which does not involve the hypothesis of an expanding universe, yet which would have avoided the paradox. This is the static universe of finite age which was discussed in section 2.

6. *Numerical Estimates*

The results of models I and II yield a total energy rate received at the surface of the Earth per unit area

$$I_{exp} = k \frac{c}{H} LN,$$

where N is the average number of radiating stars per cubic centimeter, L is the average emission rate per star, H is Hubble's constant, and k is a number. The latter has the values

$$k = \begin{cases} \frac{1}{3} \text{ (model I)} \\ \frac{1}{2} \left[1 - \frac{2}{3} \left(\frac{v_{OE}}{w_{LE}} \right)^2 \right] \text{(model IIC)} \end{cases}$$

We shall take $k \sim 1/3$ for the following discussion.

To estimate I_{exp} one should read *galaxy* where the word *star* has been used above. Then

$$N \sim 3 \times 10^{-74} \text{ cm}^{-3},$$

corresponding to a distance of 1 Mpc between galaxies. Also

$$H \sim \frac{1}{3} \times 10^{-17} \text{ sec}^{-1},$$

corresponding to a value of about 100 km sec^{-1} Mpc^{-1}. Also

$$L \sim 10^{41} \text{ to } 10^{44} \text{ ergs sec}^{-1}.$$

It follows that

$$I_{exp} \sim 10^{-2} \text{ to } 10^{-5} \text{ erg sec}^{-1} \text{ cm}^{-2},$$

which is not an unreasonable estimate for the night sky. An observational upper limit equivalent to $I_{exp} < 5 \times 10^{-4}$ erg sec^{-1} cm^{-2} was given by F. E. Roach and L. L. Smith (1968).

7. *A Note on Special Relativity*

Anyone encountering special relativity for the first time, and already familiar with Newtonian mechanics, would be forgiven for thinking that the effect of relativity is to complicate and obscure the outline of problems. In the present discussion, however, special relativity effects a great simplification. All the models of the light propagation listed in section 3b are replaced by one model: there is no ether; light is emitted isotropically in the source's reference frame; and the translation to the observer's reference frame is made by the standard Lorentz transformation. The various correction factors of section 4 are simply

$$f_D = f_a = \alpha^2, f_d = \alpha^{-1}, \qquad (7.1)$$

with

$$\alpha \equiv \left[\frac{\left(1 - \dfrac{v}{c}\right)}{\left(1 + \dfrac{v}{c}\right)} \right]^{1/2}. \qquad (7.2)$$

The expressions for f_d, including expression (7.1), have not to our knowledge been given before.*

To see the Newtonian results as a special case of expression (7.1), it is convenient to write w for the speed of light propagation *in vacuo*, and to keep c for the quantity that appears in the Lorentz transformation. Then equation (7.2) may be written

$$\alpha = \frac{\left[1 - \left(\dfrac{v}{c}\right)^2\right]^{1/2}}{\left(1 + \dfrac{v}{w}\right)}. \qquad (7.3)$$

The Newtonian limit arises for $c \longrightarrow \infty$, in agreement with the results of earlier sections.

8. *Conclusions*

In the present analysis of what Olbers might have said, we have touched on two radically different ways whereby he might have re-

---

*The Doppler effect factor is here given as α^2 because the ratio $v/v' = \alpha$ affects the received energy in two ways: the number of photons received in unit time is multiplied by the factor α, and so also is the energy $E = h\nu$ of each photon. The latter relationship could also be used in the Newtonian framework of section 4, but this was thought to be too anachronistic.

TABLE 7.1. Models of Universe

Model	Universe expands?	Paradox resolved?	Horizon exists?
I. Particle theory of light. Speed of light constant relative to source, so that light from distant sources receding faster than light does not reach the observer O.	Yes	Yes	Yes
IIA, IIB. Wave theory of light with a stationary ether. Speed of light constant, and emission of light isotropic, relative to the ether. Light from arbitrarily distant sources can reach O. The aberration factor f_a does not decrease with distance.	Yes	No	No
IIC, IID. Wave theory of light with a stationary ether. Speed of light constant relative to the ether, but emission isotropic relative to the source. Light from arbitrarily distant sources can reach O, but is weakened by a distance-dependent aberration factor f_a.	Yes	Yes	No
III. Wave theory of light with an ether participating in the expansion of the universe. Speed of light constant relative to the ether. Light from sources at distance $r > c/H$ experiences an "ether headwind" of magnitude $Hr > c$ and never reaches O. (This model has not been discussed in detail).	Yes	Yes	Yes
Static universes ⎧ of finite radius R but infinite age; $I = LNR$	No	Yes	Yes
⎨ of infinite extent but finite age T; $I = LNcT$ and increases with time	No	Yes	Yes
⎩ of infinite extent and infinite age (Olbers's model)	No	No	No

NOTE: All models are of uniform density in space.

solved his paradox:

(i) By postulating a static universe of finite age (see equation (2.11));

(ii) By postulating a universe in expansion, with continuous creation of matter so as to obtain a steady state.

These latter theories fall into two categories by virtue of different assumptions that may be made with regard to light propagation. In one category (models I, IIC, IID, III) the paradox might have been resolved. For models in the other category (models IIA and IIB) the paradox would not have been resolved; had Olbers adopted this latter group of assumptions he might well have rejected the expansion hypothesis. The two propositions P_1: "the universe is expanding" and P_2: "Olbers's paradox is resolved" are in fact logically independent; all four combinations of truth and falsehood for P_1 and P_2 are possible (and correspond to plausible models of the universe) given the knowledge of Olbers's time. The situation is summarized in Table 7.1, which also indicates under what conditions there exists a "horizon," that is a limit to the extent of the observable universe.

Suppose Olbers had seen the error in his attempted resolution of the paradox by the hypothesis of interstellar dust, as discussed in section 2. Suppose also that he decided to make a second attempt at resolving the paradox. Which of these three broad lines of reasoning would he have chosen? Indeed, would he have chosen any of them? Our conclusion in this respect is bound to be somewhat guarded, as this is a matter bordering on psychology. However, we consider that, in the absence of any direct evidence for expansion, method (i) would have been his most probable choice, since the emerging notion of energy conservation requires a finite age for each star and, therefore (barring continuous creation), for the whole Universe. The expansion hypothesis, method (ii), would have been a very inspired guess, which he just possibly might have made.

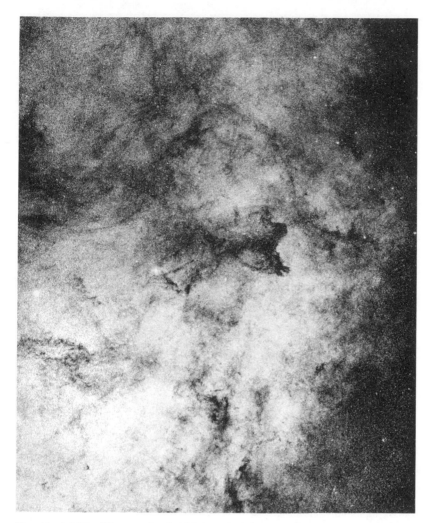

Fig. 11. A Milky Way star cloud. This is one of the most dense clouds of stars near the center of the Milky Way Galaxy in the constellation Sagittarius. The dark rifts are due to interstellar clouds of obscuring gas and dust. (Hale Observatories photograph)

Fig. 12. The giant spiral galaxy in Andromeda. This galaxy, in the northern constellation Andromeda, is our nearest major extragalactic neighbor. Our own Milky Way
Galaxy would look much the same if viewed from a distance. Note the two small
elliptical companions of this galaxy. (Hale Observatories photograph)

Fig. 13. A distant spiral galaxy, NGC 6946. Here we see the characteristic spiral arms almost face-on. All the individual stars belong to our own Galaxy. (U.S. Naval Observatory photograph)

Fig. 14. The Sombrero galaxy, NGC 4594. Another spiral galaxy; but viewed almost edge-on this time. Note the large galactic bulge-and-halo, and the dust lanes defining the galactic plane. (U.S. Naval Observatory photograph)

Fig. 15. Explosion of a galaxy, M 82. In red hydrogen light, we see gas ejected 10,000 light-years on each side of the center of this peculiar galaxy (NGC 3034). Apparently, an event of extraordinary violence took place in its nucleus some millions of years ago. (Hale Observatories photograph)

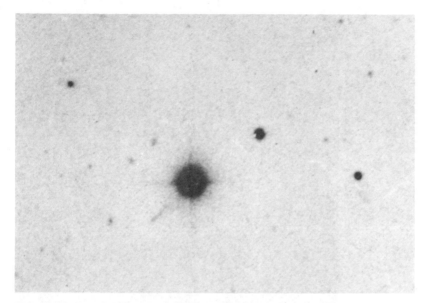

Fig. 16. First quasar discovered, 3C 273. Except for the faint jet (*lower left*), this appears to be a star. However, this quasi-stellar object is a strong radio and X-ray source, and is over one billion light-years away. (Photograph reprinted by permission from J. L. Greenstein and M. Schmidt, *Astrophysical Journal* 140: 1, 1964, © 1964 by The University of Chicago).

Fig. 17. Disk of our Sun. The first three insets show the entire solar disk in (*A*) ordinary light, (*B*) red hydrogen (Hα) light, and (*C*) calcium light. Note the sunspots and darkening at the limb of the Sun in *A*, the dark filaments in *B*, and the mottling and brightenings

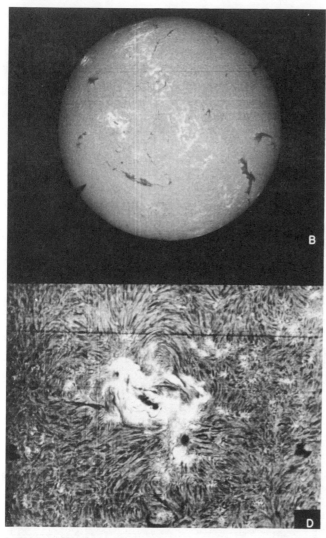

in *C*. Inset *D* is an enlargement of the sunspot group in *B*, showing the great currents and whirlpools of flowing gas on the Sun's surface. (Hale Observatories photographs)

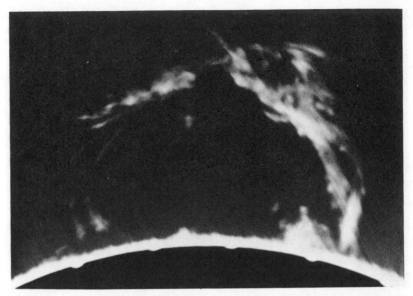

Fig. 18. A solar prominence. This arching stream of hot ionized gases more than 25 Earth diameters above the surface of the Sun. It was seen in calcium light in July 1957. (Hale Observatories photograph)

Fig. 19. A ball of stars, the globular cluster M 13. About 10 billion years ago, this group of a half-million stars formed. The globular cluster (NGC 6205) is located about 20,000 light-years from us, in our Galaxy. (Crossley photograph from Lick Observatory)

Fig. 20. A young open cluster, the Pleiades. This famous loose group of some scores of stars, imbedded in gas (see the reflection nebulosities), is located in the constellation Taurus (The Bull). (U.S. Naval Observatory photograph)

Fig. 21. Four normal galaxies. *Upper left:* the barred spiral (SBc) NGC 7741. *Upper right:* the ordinary spiral (Sa) NGC 3031 (M 81). *Lower left:* the elliptical (E0) NGC 4486 (M 87). See the lower left portion of Fig. 22 for a (12×) short exposure showing the jet in the central region of this giant galaxy. *Lower right:* the ordinary spiral (Sb) NGC 4565 seen edge-on, with a dust lane silhouetted against the central bulge. (Hale Observatories photographs)

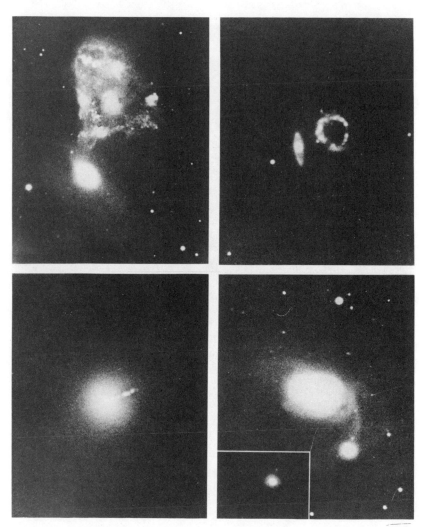

Fig. 22. Four peculiar galaxies. *Upper left:* the peculiar galaxies NGC 2444/2445 (Arp 143). *Upper right:* the peculiar galaxy Arp 147. *Lower left:* the jet in the center of NGC 4486 (M 87; shown in long exposure in Fig. 21). *Lower right:* a long-exposure 200-inch photograph of NGC 7603 by Arp. A luminous bridge appears to join the galaxy (whose red shift corresponds to 8800 km sec^{-1}) to its companion (16,900 km sec^{-1}; see the companion in detail in the inset). (First three photographs reprinted by permission of the author from H. Arp, *Atlas of Peculiar Galaxies*, 1966; last photograph reprinted by permission from H. Arp, *Astrophysical Letters* 7: 221, 1971, © 1971 by Gordon and Breach Science Publishers Ltd.)

Fig. 23. Center of the Leo group of galaxies. These are the four brightest members of the Leo group. The elliptical (E2) NGC 3193 is at *upper left*, and the barred spiral (SBa) NGC 3185 at *lower right*. The two central spiral galaxies, NGC 3190 (Sb; *lower center*) and NGC 3187 (SBc; *upper center*), show signs of strong tidal distortion. (Hale Observatories photograph)

8. Big-Bang Cosmology: The Evolution of the Universe

George B. Field

University of California, Berkeley

The words *big bang* connote explosion. The word *cosmology*, from the Greek *kosmos logos*, denotes the meaning of the Universe. Can the Universe, as the title of this essay implies, be exploding and evolving? Our intent here is to explore in detail a "standard" model of our Universe, the so-called big-bang cosmology. Along the way, we will meet alternative viewpoints, which indicate that we have not yet attained certain knowledge in our age-old quest for a consistent world-view.

Timeless Stability

We human beings tend to perceive the physical world as unchanging, but for the welcome rhythms of day and night, and the weather. Indeed, since its inception mankind has striven to limit, control, counteract, and even disregard the evident flux of Nature: We created religion and mysticism to "explain away" Nature in anthropomorphic terms; we constructed society so that our surroundings and interrelationships with one another could become more "understandable" (*i.e.*, artificial); we learned to speak and think in conventional, linear, sequential patterns of logic, further constraining reality to our mold. Our struggle for stability amidst apparent chaos has been marked by epochs of success (unkindly termed "stagnation" by some historians), but Nature and human character always reemerge in revolutionary ways. The present, "modern" era is an example, for the all-too-evident changes of Nature have once again forcefully intruded themselves into our conservative thought patterns.

How has it been possible for us to maintain a belief in the timeless stability of our Universe? The fundamental answer is this, that we find ourselves in an almost-unique environment. We are short-lived creatures, exceptionally influenced by our solid-state surroundings. The heavens (and even the Earth) appear immutable, because almost all cosmic phenomena operate on time scales that vastly exceed the span of our existence; our limited senses cannot easily decipher cosmic changes, and our poor memories obviate many of the changes that we do see. We exist in a primarily solid world, wherein we can use nouns to label "objects"; tables and rocks appear stable during a lifetime, but of course, they inevitably change on long (*e.g.*, geological) time scales. The circumstances and exigencies of our existence have caused us to be static parochialists.

Timely Change

But all of that is now changing! The monolithic Newtonian structure of science and the relative stability of society that characterized the nineteenth century have given way to the uncertain, relative, and polymorphic manifestations of evolution in the twentieth century. Let us attempt to see what these changes are all about.

After circumnavigating the Earth aboard the *Beagle*, Charles Darwin had enough evidence virtually to prove that life evolves over the eons from simpler to more complex forms. Today, on the basis of the Watson-Crick double-helix model of DNA, we understand that mutations (*i.e.*, random changes) lead to this evolution—under the pressure of the struggle for existence—by altering the A-T-G-C base sequence of the genetic code (see Essay 5). Even *homo sapiens* is subject to this relentless process; witness the extinction of Neanderthal man.

Our seemingly stable Earth (see Fig. 3) is really not so stable after all. The oceans and rivers nibble away at the land, while vast mountain ranges rise slowly from the plains. If we could compress a thousand years into a second, we would see the Earth's surface writhing and contorting. Continents would change shape rapidly as the oceans rise and fall; the vast glacial blankets of the ice ages would swiftly advance from near the poles and then recede again; the Earth would be spinning like fury and precessing like a child's top; our Moon would be seen to recede slowly from the Earth as a result of tidal friction and energy losses; and the primeval continental masses would be rent asunder, the pieces drifting over the surface, plowing into one another (creating the Himalayas where India ran into the southern end of Asia), and spouting chains of volcanoes at their leading edges. This last feature has been proved only within the last few decades; the *continental drift*, at the rate of inches per year, began about 200 million years ago and is proceeding even today. This movement drastically influenced organic evolution, as plant and animal populations were separated by the ever-drifting continental blocks.

On an even longer time scale, we find the planets of our Solar System forming from the interstellar medium (along with our Sun), about 4.6 billion years ago. Detailed studies of the remnant radioactivity in rocks of the Earth's crust, in samples of the Moon's surface, and in meteorites point to their essentially simultaneous formation at that epoch. About 3.5 billion years ago, an event of extraordinary violence seems to have occurred in the Earth-Moon system (the capture of the Moon by the Earth), for at that time it appears that (a) the surface of the Earth melted and slightly later life appeared, and (b) the dark lava-filled maria on the Moon's nearside were formed. On a scale of billions of years, our Solar System has been anything but static.

What about the multitudinous stars that constitute our Milky Way Galaxy and the billions of other galaxies that we see in space? As Essays 1, 2, 3, and 5 have shown, a star is not eternal but is rather a dynamically evolving entity. With its finite store of gravitational and thermonuclear energy our Sun (Fig. 17), for example, can last only about 10 billion years; then it must fade away into a cinder-like black dwarf. If new stars were not continuously being formed from the interstellar medium, as they clearly are in spiral galaxies (see Essay 6), only dim, conservative stars of less than two solar masses would remain to beckon us from the celestial sphere. Clearly, even galaxies are "living, breathing" entities. (Note, especially, the active and "exploding" galaxies mentioned in Essay 6.)

The galaxies tend to aggregate into groups and clusters of galaxies (see the frontispiece and Fig. 23), and these clusters appear to fill cosmic space uniformly to the limit of detectability—about 10 billion light-years. Scattered here and there are the brilliant, hyperactive quasars (see Fig. 16), some of which are almost at the "known edge" of the Universe (C. R. Lynds ascribes a red shift of 2.877 to the quasar 4C05.34; this object is receding from us at almost 90 percent of the speed of light.). M. Schmidt has found that, the farther out we look (*i.e.*, the greater the red shift), the more dense becomes the population of quasars, that is, their number per unit spatial volume; it therefore appears that even the Universe itself is evolving.

Cosmic Evolution: The Observations

What basis, besides philosophy and metaphysics, do we have for believing in the evolution of our Universe? Let us permit the observations to tell their tale.

Not only does the number density of quasars appear to be increasing as we look into the past (*i.e.*, to great distances; remember that their light can only travel at the finite speed of 186,000 miles per second, so that we now see them as they were when they emitted the radiation), but radio source counts (see Essay 10) seem to show evolutionary effects at great distances (that is, for the apparently weakest sources).

The modern theory of stellar structure and evolution has enjoyed tremendous success in explaining the H-R diagrams (see Essay 1) of stars and stellar clusters. We find that we can deduce a reliable age for a cluster from its main-sequence turnoff point on such a diagram. The oldest stars and clusters in our Galaxy have an age of about 10 billion years. By extrapolating the well-known Hubble expansion of the Universe backwards in time, we arrive at the remarkable result that all galaxies would have been at one and the same point about 10 billion years ago. We term this the "beginning" of the Universe. That these two independently determined numbers should be even close to the same must be more than a chance coincidence.

Let us recall the form and content of Edwin Hubble's *law of the red shifts*. In the 1920's Hubble and his colleagues discovered that the spectral features from other galaxies were displaced toward longer wavelengths (*i.e.*, red shifted) relative to the same spectral features in laboratories on Earth. In addition, for a given type of galaxy, the red shift was greater the dimmer the galaxy appeared. By assuming that the galaxies have the same intrinsic brightness, we deduce that they appear dimmer the more distant they are. By interpreting the red shifts as a Doppler-velocity effect, *i.e.*, caused by velocities of recession, Hubble found that *the galaxies are receding at speeds proportional to their distances from us*. Today, largely through the efforts of A. Sandage, this proportionality has been verified from a few thousandths up to half the speed of light.*

---

*Since all mass-energy "attracts" (even in general relativity), the universal expansion must slow down, or decelerate. At large red shifts, beyond about 1, this effect appears in the Hubble relation as a deviation from the straight line of strict proportionality. We parameterize this deviation with the *deceleration parameter*, q_o, which (essentially) measures the rate of slowing-down in terms of Hubble's constant. To date, Sandage finds $q_o = 0.9 \pm 0.4$.

The constant of proportionality—the Hubble constant—has the unit of inverse time; its evaluation implies the characteristic expansion time (from a point) of 10 ± 2 billion years. It appears that the galaxies formed approximately that long ago, when the Universe was extremely small and dense.

George Gamow accepted the ultimate extrapolation of the Hubble expansion back to the infinitely dense, singular "beginning"; using well-known physical laws, he predicted that this incredibly hot initial state should be a fireball of photons. We can still see the fireball in the distant past at a great linear distance, since these initial photons must spend 10 billion years traveling to us; that is, they must appear to originate 10 billion light-years away. But this "edge" of the Universe, which is also its temporal beginning, is receding at almost the speed of light; hence, the relict fireball photons are red shifted by a factor of 1000 before they can reach us, and they must therefore appear rather "cold." In 1965 A. Penzias and R. Wilson of the Bell Telephone Laboratories discovered a strange *background microwave radiation* coming from space; it turns out to be *isotropic* (arriving with incredible uniformity from every direction) and has a blackbody temperature of only 2.7 degrees absolute. This radiation has now been "seen" throughout the wavelength range from 1 millimeter to 21 centimeters, where it defines a blackbody curve at 2.7 degrees; we have apparently found the 3000-degree fireball red shifted 1000 times, as predicted by Gamow!* If these observations withstand the test of time, then our Universe is indeed evolving and big bang.

Finally, there is cosmic nucleosynthesis, and the abundances of elements in the Universe. One implication of Gamow's "hot beginning" is that as the Universe expanded and cooled, approximately 30 percent helium (by mass) was formed from the initial hydrogen at temperatures around 10^9 degrees in the first half hour of the Universe. When we look about us, we find that both stars and galaxies (see Essay 6) have at least this minimum helium abundance,† as they must if the theory is correct. This is another striking verification of the evolving, big-bang picture of our Universe.

---

*Recently, rocket infrared observations have revealed anomalous radiation at wavelengths below 1 millimeter; but the correct interpretation of these data is still an open question. These results, if correct, threaten the "primeval fireball" picture.

†There are a few peculiar stars which appear to have almost no helium; but since we can see only their luminous atmospheres, we have no way (at present) of knowing whether the helium has settled out of the atmosphere, or the entire star is helium-deficient.

Cosmological Models Galore

Before explicating big-bang cosmology, let us briefly consider the plethora of theoretical cosmological models that presently exist.

The preceding essay dealt with Newtonian cosmology in some detail. Since Isaac Newton's (1688) *law of universal gravitation* implies that every mass attracts every other mass, all Newtonian models must be dynamical (*i.e.*, expanding, to agree with the observations). Hence, a Newtonian universe is necessarily big-bang, and in fact, the Newtonian picture is an excellent approximation to the general relativistic model. But Solar System observations show that Newton's theory cannot be completely correct, for its fails to account for (a) the anomalous 43 arc-seconds per century precession of the perihelion of the planet Mercury (the point of Mercury's orbit that is closest to the Sun moves this angular distance every century), (b) the 1.75 arc-seconds bending of light rays as they pass near the edge of the Sun, (c) the red shift of light as it climbs out of the Sun's gravitational clutches toward the Earth, and (d) the well-verified effects of special relativity.

Albert Einstein's (1916) *general theory of relativity* will serve as our basis for discussing big-bang cosmologies in the final section of this essay. In this theory the material content of the Universe warps the structure of space and time, and that structure in turn determines the dynamical motion of the material content. This theory satisfactorily accounts for the four effects that eliminated Newtonian gravitation from our consideration. Cosmological models based upon Einstein's equations are inevitably of the big-bang variety,* and they appear to account for all the presently known observations.

C. Brans and R. Dicke have advanced a variant of general relativity in their *scalar-tensor theory of gravitation.* Here an *ad hoc* scalar field is appended to the usual tensor field of general relativity; the effect is to alter the time scale of big-bang, evolutionary cosmologies but not to render them unrecognizable. At present, the observational data argue against this complicated theory, but it will be a few years before it can be conclusively tested.

Finally, the only cosmological theory that is still a serious contender with general relativity is H. Bondi and F. Hoyle's *steady-state theory* (circa 1948). (Essays 7 and 10 comment upon this theory.) Ac-

---

*We will completely disregard the so-called *cosmological constant* λ. It seems to represent a nonphysical force, and Einstein expressed regret that he had introduced it.

cording to this theory the Universe expands but on the average remains unchanged in both space and time; hence, new matter must
appear from nowhere to fill the gaps left by the receding galaxies,
and this new matter must form galaxies that also partake in the
cosmic Hubble expansion. Since the new matter is created at such a
low rate that the process can't be detected in the laboratory, the
principal appeal of this theory is its philosophical "beauty." Unfortunately for their creators, steady-state cosmological models run
into observational difficulties: (a) since they are nonevolutionary,
they cannot explain the quasar and radio source density behaviors;
(b) since they are non-big-bang, they cannot explain the universal
helium abundance nor the cosmic background radiation; and (c)
they predict that we should see large numbers of newly forming or
young galaxies, which we do not.

The "Standard" Big-Bang Scenario

We have seen that everything in our Universe evolves—nobody dis-
putes this—so let us consider the big-bang cosmology of general relativity, where our entire Universe also evolves—some dispute this.

In the beginning, our Universe "appeared" with zero volume, infinite density, and infinite temperature; known physical laws cannot
really describe this initial singular state (but see Essay 9). Within a
millisecond (one thousandth of a second), the temperature of the
holocaust dropped to a trillion degrees due to the rapid universal
expansion, and the known elementary particles (proton, neutron,
electron, and so on) began to precipitate out. A few minutes later,
the matter was cool enough, about a billion degrees, for thermonuclear reactions to form helium nuclei from protons and neutrons,
the so-called *cosmic nucleosynthesis*. Much smaller amounts of heavy
hydrogen (H^2) and light helium (He^3), but totally negligible amounts
of heavier nuclei (lithium and higher) were formed at this stage. For
the next 100,000 years, the Universe (and its matter content) was
dominated by the intense fireball of photons; but at the end of this
period the ions became cool enough to form neutral atoms (of
hydrogen and helium), and the photons decoupled from the matter
at about 3000 degrees (we see these photons today as the highly red-
shifted cosmic microwave background). Small density fluctuations in
the cooling gas began to self-gravitate and grow, until they over-

came the cosmic expansion and collapsed into proto-galaxies; this galaxy formation occurred about 100 million years after the initial big-bang event.

Consider our own Galaxy. About 10 billion years ago, it was a tenuous, almost-spherical cloud of turbulent gas beginning to contract out of the cosmic background. Its radius was about 100,000 light-years, and its temperature only a few degrees absolute. The collapse occurred rapidly, on a time scale of about 200 million years, during which time the metal-deficient Population II stars and globular clusters (see Fig. 19) formed in the galactic halo. As the collapse continued, rapid star formation, and the sudden deaths of massive stars (and other supermassive objects), led to a metal-enrichment of the galactic environment; soon, newly forming stars were metal-rich, that is, of the Population I category. The initial turbulence of the gas began to manifest itself as an overall galactic rotation during the final phases of collapse to the galactic plane. The end result was the tenuous spheroidal halo, enveloping the large central bulge of stars, with the thin, dense galactic disk-and-plane bisecting the whole ensemble. Nonuniformities in the rotating disk triggered *spiral density waves* (traveling gravitational patterns that maintain themselves by the slightly greater density of the materials that they attract). Disk gas-and-dust clouds, catching up with these "standing waves," are rapidly compressed into a dense shock wave, from which new stars may form. It is these bright, young (Population I) stars that reveal the density waves as spiral arms in the Galaxy (see Figs. 12 and 13).

About 5 billion years after the formation of our Galaxy, our small Sun condensed from the interstellar gas and dust in a spiral arm about 30,000 light-years from the center of the Galaxy. It appears that the planetary system accreted from the residue of this solar formation 4.6 billion years ago; the Earth was born. Approximately 1 to 2 billion years later, organic life arose on the surface of the ever-changing Earth, and evolutionary processes eventually led to complex life-forms—like ourselves.

The Future

Our theory seems to have far outdistanced our meager facts, but we must be careful. We have elucidated the "standard" big-bang cosmological model based upon general relativity. E. Harrison refers to this model as one of the "grin cosmologies," for it is stripped of all complications (such as inhomogeneity, actual turbulence, real galaxies, and so on) and deals only with the gross structure of universal evolution. In fact, we run into almost insurmountable difficulties when we actually attempt to produce galaxies from the "smoothed-out" cosmic materials. We have essentially eliminated the entire Cheshire cat and are left with only his enigmatic *grin*.

Of the fast-fading Cheshire cat (in Lewis Carroll's *Alice's Adventures in Wonderland*, 1865) only the grin ultimately remains. This picture serves as a warning for us not to eliminate all of the essentials from our cosmological models of the Universe, for we may end by "throwing the baby out with the bath water." In other words, never underestimate (or undermodel) the true complexity of Nature.

Hence, the program for the coming years must consist of (at least) two parts: (1) we must strive to make our models less idealized and more realistic, and (2) we must expand our observational understanding of the enormous gap between the quasar red-shift limit of 3 and the cosmic-background-radiation red shift of 1000, for this is where we expect to find galaxies in the process of formation. If Nature is kind, and present indications are verified, then we may speak with great confidence of the "big-bang Universe." At that future date, we can believe that we understand the evolution from elementary particles to atomic nuclei, to atoms, to molecules, and finally to life in the expanding Universe.

Then as now, however, we will remain confronted by the ultimate enigma: what caused the explosion in the first place?

9. A Quantum Universe: The Beginning of Time

Kenneth C. Jacobs
University of Virginia

> I have a bit of FIAT in my soul,
>   And can myself create my little world.
>   —Thomas Lovell Beddoes, *Death's Jest Book*, V.i. 38

The preceding essay describes the basic properties of the standard big-bang cosmological models of our Universe. Since the discovery of the cosmic background radiation in 1965, these models have gained favor with most cosmologists, for the 2.7°K (absolute temperature) blackbody radiation from a primeval fireball follows naturally from the assumption of a hot big bang. Today we know that our Universe of galaxies is expanding faster the farther out (or further back in time) we look. To recall the essential characteristics of big-bang cosmology, let us reverse the arrow of time and watch the Universe shrink back toward its beginning about 10 billion years ago. Individual galaxies, now typically a million light-years apart, touch and merge together when the Universe has contracted by a factor of 30 or so. Going back still further, before any stars have formed, we have a nonuniform gas composed of neutral hydrogen (70 percent by mass) and helium (30 percent), which ionizes when we reach about 250,000 years from the beginning of the expansion. All these morphological processes occur during the *matter era;* but before 2000 years the energy contained in the 2.7°K blackbody radiation is so compressed that it dominates the dynamics of the Universe; we have entered the *radiation era*. The temperature of the Universe rises in proportion to the linear contraction, and at about one billion degrees Kelvin thermonuclear reactions take place (during the first few minutes of the Universe). One second from "creation" we cross into the *lepton era* (at 10 billion degrees Kelvin), where particle-antiparticle pairs of electrons, muons, and neutrinos appear in abundance. As the compression increases, and the temperature surpasses one trillion degrees (in the earliest microseconds), massive-particle pairs (*e.g.*, protons, neutrons, lambdas, sigmas, and so on) announce the threshold of the *hadron era*.

Thus far in our theorizing we have been on firm ground and could proceed with some confidence; but when cosmic densities become greater than *nuclear densities* (*i.e.*, the density of an atomic nucleus: $\rho \approx 10^{14}$ grams per cubic centimeter $\approx$ 100 million tons per teaspoonful), we encounter the unexplored no-man's-land of the first microseconds of existence. Here, within a flicker of the beginning of time, our footing becomes slippery. Does the temperature of the Universe continue to rise indefinitely (to infinity!) as we approach $t = 0$, or does it level off near a "boiling point" of $kT \approx 150$ MeV due to the production of an infinite spectrum of strongly interacting particles? Indeed, does the statistical equilibrium concept of "temperature" have any meaning under such extreme conditions? What about the matter content of the Universe, which is here such an insignificant fraction (about one billionth) of the total particle-antiparticle density? It is this relatively minute amount of matter that eventually forms all the galaxies in our Universe, but we cannot account for its origin. And finally, we may ask the ultimate metaphysical questions: Where was our Universe before it began? What happened before the beginning of time? How and why did our Universe begin?

In this essay we will concern ourselves with the beginning of the Universe, and with the first microseconds of its existence. If we can account for this epoch, perhaps the current content and evolution of our Universe will become clearer in our minds. To accomplish this task, we must enter the realm of quantum cosmology, where time, gravitation, and quantum mechanics are inseparably bound together.

Of Time and Cycles

What do we mean by the word *time*? Anthropomorphically we sense duration—psychological time—by referring to heartbeats or to electrical impulses in our brains. Time is defined operationally as the *cyclic repetition* of some standard phenomenon. Each "tick" of a pendulum clock marks the completion of a cycle of the pendulum bob's motion. Hence, cycles and time are inextricably tied together.

A pendulum clock certainly suffices for a crude measurement of time, but it is far from ideal in most cases; on the surface of the Moon a given pendulum will oscillate much more slowly than it does on the Earth's surface, and on a roller coaster the device is useless. A more trustworthy timekeeper is our rotating Earth, but in comparison to very stable atomic and molecular clocks even our planet keeps imperfect time. From considerations such as these, we find that a perfect clock is one which is totally unaffected by all external perturbations. Since our Universe is the sum total of everything, only it qualifies as a perfect clock.

Why are we so concerned about perfect clocks? Our topic is the beginning of all time; how will we know when $t = 0$ if it cannot be measured? This brings us to the crux of the matter: does time itself exist, if there is no way to measure it at the beginning? If our Universe were completely classical, that is, if quantum mechanical phenomena were not important, the answer to this question might be yes, for we could imagine a sequence of cyclic processes of ever-increasing frequency. Consider how time is kept in a big-bang cosmology: At the present epoch our Milky Way Galaxy is a good timepiece, since it rotates once every 250 million years (about forty cycles since the beginning). Before the galaxies formed, time could be counted using the atomic frequencies of hydrogen (periods of order 10^{-8} second); earlier still, the characteristic decays (about 10^{-23} second) of heavy hadrons might have sufficed. However, as we approach $t = 0$ each of these clocks "freezes out," that is, the oscillations appear to slow down and effectively stop, and a faster clock must be called into service. An unstable particle of infinite mass might decay instantaneously, thus marking the origin of time; but aren't we being excessively wishful to postulate such a particle? To specify the possible big-bang cosmologies we use Albert Einstein's general theory of relativity, but this classical theory characterizes the beginning as a *singularity*. In such a singularity, the mass-energy density (remember the famous equation $E = mc^2$) of the Universe becomes infinite at $t = 0$; hence, a "clock" is subjected to infinitely strong perturbations at $t = 0$. The resolution of these temporal paradoxes will become apparent only after we consider the quantum dynamics of the early Universe.

Quantum Gravitation

To understand the stage upon which quantum cosmology is played, we must first be conversant with gravitation and quantum mechanics.

Gravitation is a ubiquitous property of all matter (and energy); it is most familiar to us in the form of the gravity that unceremoniously holds us to the surface of this Earth. Since the late seventeenth century, when Isaac Newton propounded his law of universal gravitation ($F = GMm/r^2$), we have subconsciously attributed gravitational phenomena (the trajectory of a missile or the orbits of the planets in our Solar System) to a mutual attraction between the bodies involved. But what exactly is an attraction (other than a common-sense feeling), and how fast can such an influence propagate? If I jiggle the Earth, will the Moon in its orbit respond instantaneously? Einstein considered such questions, and between 1905 and 1916 he set forth a general theory of relativity to answer them. First, he joined the three dimensions of space and the one dimension of time into an inseparable four-dimensional entity called *space-time;* as a consequence, he found that (locally) no particle or information-bearing disturbance could propagate faster than the speed of light in vacuum ($c \approx 3 \times 10^5$ kilometers per second). For example, the Moon cannot respond to the jiggling Earth until at least $1\frac{1}{4}$ seconds have elapsed. Secondly, the geometry of space-time is "flat," until we disturb it by placing mass (or energy) nearby. In accordance with Einstein's field equations, matter warps the geometry of the surrounding space-time; by trying to pursue the "straightest" path in this curved space-time, a passing particle (or even a light ray) actually traces out a curved trajectory in three-dimensional space (*e.g.*, the planetary orbits about our Sun). Einstein's theory subsumes Newton's, and in addition predicts phenomena that are well-verified observationally (both in terrestrial laboratories and in the Solar System). Though classical, in the sense that it takes no account of quantum mechanics, the new theory of space-time geometry is philosophically and conceptually far superior to its predecessors.

Quantum mechanics (or wave mechanics) is the theoretical edifice raised in the 1920's and 1930's to account for the strange (and apparently contradictory) behavior of particles in the atomic domain (regions smaller than about 1 Angstrom = 10^{-8} centimeter). Here Newton's laws of mechanics no longer hold, for light waves are seen to act like "solid" particles (*photons*) and seemingly solid atoms diffract and interfere in a most wavelike fashion. The new quantum dynamics describes the fact that every entity can manifest itself either as a wave (in terms of probability and energy) or as a particle (in terms of conserved properties and equivalent mass). At one time the combination term *wavicle* was in vogue to characterize these entities.

Quantum theory, especially in conjunction with special relativity, has several implications of great import to our present discussion. The Heisenberg *uncertainty principle* tells us that the accuracies to which we can specify both the position (Δx) and the *momentum** (Δp) of a particle (simultaneously) are constrained by the limit $(\Delta x)(\Delta p) \gtrsim \hbar/2$, where $\hbar$ is about 10^{-27} erg second, an exceedingly small quantity. The significance of $\hbar$ in terms of angular momentum was described in Essay 1. As a consequence of this principle, even the most stable state of a quantum system, the *ground state*, is characterized by irreducible quantum fluctuations in energy. An equivalent formulation of this principle, related directly to the wave nature of all quantum phenomena, states that the uncertainty in the energy of a system (ΔE) is related to the lifetime (Δt) of that system by $(\Delta E)(\Delta t) \gtrsim \hbar$.

Using the uncertainty principle, quantum mechanics can account for the *electromagnetic interaction* in terms of the "virtual" (*i.e.*, energy nonconserving, and unobservable) exchange of photons between charged bodies; similarly, the *strong interaction* between hadrons is transmitted by "virtual" massive mesons of various types. It is only natural, therefore, to ask, What mediates the *gravitational interaction*? It is possible to specify a massless (spin 2) particle which can perform this task: the *graviton*. Hence, it would appear that we have succeeded in uniting gravitation and quantum mechanics; in the ultraweak-field limit, this is true, for we can reproduce nonrelativistic gravity (*i.e.*, Newton's theory). By allowing the graviton to interact with other gravitons and going to the nonlinear classical limit of large numbers of gravitons, we can derive Einstein's representation of geometrical gravitation. But we have not fully quantized the gravitational field, for all the anticipated quantum fluctuation phenomena are still absent. Perhaps it would be better to proceed on a different tack, as J. A. Wheeler has done.

Before we launch into quantum cosmology as such, let us first enumerate some useful consequences of relativistic quantum theory. As P. A. M. Dirac was the first to show, for every *particle* (*e.g.*, the electron, symbolized as e^-) there is associated an *antiparticle* (the positron, e^+); a particle-antiparticle pair can annihilate and their mass changes into pure energy ($e^+ + e^- \rightarrow 2$ photons). Moreover, by concentrating enough pure energy, we can also create a massive pair

---

*Momentum is the product of a particle's mass and its velocity, so that it characterizes the particle's "impetus of motion."

consisting of a particle and its antiparticle from the vacuum (*e.g.*, photon + atom $\rightarrow$ atom + e^+ + e^-). The spin-statistics theorem states that there are only two types of particles: *fermions* with half-integral spin (*e.g.*, electron, muon, neutrino, proton, sigma, and so on), and *bosons* with integral spin (photon, pion, graviton, eta-meson, and so on). Essay 1 has discussed some of the phase-space properties of fermions (*e.g.*, degeneracy); bosons differ in that any number of their kind can be fit into a cell of volume $(2\pi\hbar)^3$ in the six-dimensional phase space, so that a phenomenon called *Bose condensation* becomes possible at low enough temperatures. The Pauli exclusion principle applies only to fermions.

The Stage is Superspace

Today quantum mechanics embraces every domain except that of cosmological gravitation. Since our context is Einstein's general theory of relativity—a theory of space-time geometries—we seek to quantize geometry. The geometry of curved space-time is specified by a *metric*, that is, a mathematical expression which tells us the "distance" between two neighboring points in the space-time. When no mass-energy is present, we have the familiar flat (Minkowski) metric of special relativity:

$$ds^2 = (dt^2) - (dx^2 + dy^2 + dz^2), \tag{1}$$

where ds is the interval between the point (t, x, y, z) and the nearby point $(t + dt, x + dx, y + dy, z + dz)$. Note in equation (1) that the second term on the right is the Euclidean spatial metric, familiar to us from the Pythagorean theorem for the linear distance in three-dimensional space. The simplest metric of big-bang cosmology is related to equation (1), but with a time-dependent scaling, $R(t)$, of the spatial metric to account for the expansion of the Universe:

$$ds^2 = (dt^2) - R^2(t)(dx^2 + dy^2 + dz^2). \tag{2}$$

The cosmological model of equation (2) is defined by the temporal evolution of the spatial (*i.e.*, 3-space) metric, so we must know $R(t)$. The behavior of $R(t)$ follows directly from Einstein's field equations, as a consequence of the assumed material content of the cosmology.

In 1968 J. A. Wheeler at Princeton University invented *super-*

space, the arena inhabited by all of the 3-metrics consistent with the Einstein field equations. In analogy to the six-dimensional phase space (of position and momentum) mentioned in Essay 1, we may visualize superspace as an infinite-dimensional array of points, with each point representing one spatial metric (*e.g.*, the metric in equation (1)). In a static or stationary case, such as totally empty space-time or the spherically symmetric (Schwarzschild) space-time surrounding a star, this single point in superspace summarizes the entire metric. But in a dynamical situation, such as a big-bang cosmology, the spatial metric is time-dependent, and so there is a different superspace point for every value of *t*. Hence, as displayed in Diagram 9.1,

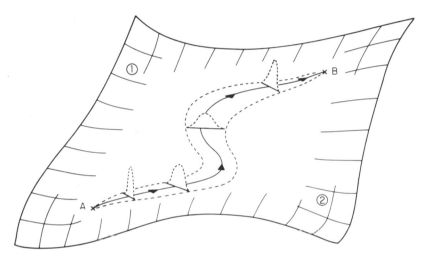

Diagram 9.1. Wheeler's superspace arena is an infinity of 3-metrics, represented here by a curved surface. Points (1) and (2) are static space-times, while the solid curve *A B* is an evolving classical cosmology. A probability wavepacket (dashed and shaded regions) arises when we quantize the geometry.

the evolution of a cosmological model is equivalent to a trajectory in superspace. Armed with these crucial concepts, we can now attempt to quantize geometry.

   The superspace path shown in Diagram 9.1 corresponds to the history of a classical cosmology, and so it is similar to the deterministic trajectory of a particle obeying Newtonian dynamics in ordinary space. In the usual transition to quantum mechanics, precise trajectories are replaced by indistinct probability waves, which tell us only the likelihood that we might be following a particular path. Generalizing to superspace (see the dashed lines in Diagram 9.1), we supplant the well-defined classical Einstein path with a quantum probability packet moving through superspace. To obtain quantum gravitation, and thence quantum cosmology, by such a generalization is exceedingly difficult mathematically; therefore, many different approaches have been pursued. One of the more fruitful efforts considers only highly symmetric cosmologies, that is, those which are *spatially homogeneous* (at a given cosmic time, every fundamental observer sees exactly the same thing) and which have some *geometrical symmetry* (such as isotropy, axial-symmetry, or translation-invariance). These conditions reduce infinite-dimensional superspace to a two-dimensional minisuperspace, and quantization now involves (a) assuming an appropriate metric, (b) postulating the material content, (c) summing the kinetic, potential, and space-time-curvature energies in the form of a Hamiltonian function, (d) making the standard quantum substitutions into this Hamiltonian, and (e) solving the resultant equations for the evolution of the probability waves in minisuperspace.

One fascinating example of such a quantum cosmology is the "anisotropic-closed" case, generically known as Bianchi type IX: this big-bang model is spatially homogeneous, is not isotropic (that is, the view is not the same in every direction), does not rotate, and contains enough mass-energy content to reverse its initial expansion so that future collapse to a second singularity inevitably occurs. Diagram 9.2 (left) depicts the classical evolution of such a model. Note the tortuous trajectory in superspace, caused by repeated reflections of the superspace point from a nonstatic potential well (right). The oscillations of this path in the $(\beta_+, \beta_-, \Omega)$ minisuperspace represent nonuniform expansions and contractions of a typical volume in the model. Quantization of the geometry yields discrete states of anisotropy, analogous to atomic energy levels, between which transitions may occur as this cosmology evolves (*e.g.*, excitation and de-excitation). To reproduce the classical history, we simply superpose many quantum states to form a wave packet which follows the classical trajectory in minisuperspace. An interesting consequence of this procedure is that, while the wave packet spreads somewhat with each bounce from the potential wall, such a packet remains essentially classical; the relative sizes of the packet and the well remain constant when averaged over several bounces.

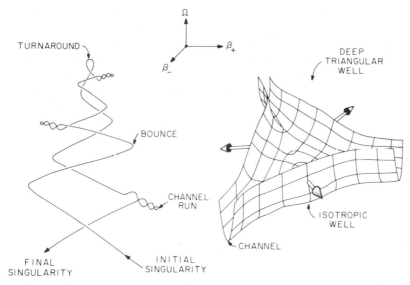

Diagram 9.2. The classical evolution of a Bianchi type IX cosmology ("anisotropic-closed" model) is shown by the superspace path (*left*); the expanding, triangular potential well (*right*) is responsible for this trajectory. The classical superspace point bounces around within the well, while discrete states arise when the cosmology is quantized.

Quantum Foam

Where are the quantum fluctuations in our superspace picture? From simple dimensional arguments, we can say the following: In quantum cosmology, we have three physical constants to work with, (1) the gravitational constant, $G = 6.670 \times 10^{-8}$ cm^3 g^{-1} sec^{-2}; (2) the speed of light, $c = 2.998 \times 10^{10}$ cm sec^{-1}; and (3) Planck's quantum of action, $h = 6.626 \times 10^{-27}$ cm^2 g sec^{-1} (note that $\hbar = h/2\pi$). From these constants, we can construct the following characteristic quantities:

(a) a *length* $= (G\hbar/c^3)^{1/2} = 1.62 \times 10^{-33}$ cm;
(b) a *time* $= (G\hbar/c^5)^{1/2} = 5.39 \times 10^{-44}$ sec;
(c) a *mass* $= (\hbar c/G)^{1/2} = 2.18 \times 10^{-5}$ g; and
(d) a *density* (mass/volume) $= c^5/G^2\hbar = 5.16 \times 10^{93}$ g cm^{-3}.

From such numerology, we expect that quantum gravitational fluctuations should appear at densities like 10^{93} g cm^{-3}, in a size of about

10^{-33} cm, with a time scale near 10^{-43} sec. Wheeler has given these fluctuations the name *quantum foam;* he envisages a chaotic sea of such ripples throughout the geometry of space-time, with the familiar fundamental particles (*e.g.*, electron, proton) being mere "clouds" propagating through the foam. Today our experimental knowledge extends only to densities as great as 10^{14} g cm^{-3} (nuclear densities), lengths as short as 10^{-15} cm (high-energy electron scattering), and times as brief as 10^{-23} sec (hadronic decays); hence, the domain of quantum gravitational fluctuations is very far removed from laboratory physics. In the cosmic "laboratory" of our Universe, however, such phenomena will assume paramount importance at the beginning of time.

Now let us focus on the "creation" of a quantum cosmology in superspace. The quantum foam appears in superspace as an omnipresent haze of evanescent geometries; in other words, this is the superspace analog of "virtual particles in the vacuum." On this scale, every geometry, including every cosmology, is subject to an inherent uncertainty (recall the Heisenberg uncertainty principle). The concept of time loses all meaning for periods shorter than 10^{-43} second, so that all paradoxes associated with $t = 0$ are meaningless (*i.e.*, the paradoxes are "resolved"). Simple calculations of the behavior of a single "bubble" of the quantum foam seem to confirm this remarkable interpretation.

In the Beginning . . .

To see where quantum cosmology leads us, we will outline a single scenario of the beginning of our Universe. Somewhere in the infinitude of superspace, a quantum geometrical fluctuation appears, endowed with rotation, anisotropy, and curvature energy. Within 10^{-43} second this "bubble" expands, so that the initial discrete states of rotation and anisotropy, feeding upon one another, decay into a plethora of gravitational inhomogeneities: gravitons. We have begun with nothing but quantum geometry. Being filled with a graviton gas (analogous to the well-known blackbody gas of photons), this fluctuation does not immediately subside back into the obscurity of the overall quantum foam, but rather continues to expand. Now graviton-graviton interactions occur and copiously produce particle-antiparticle pairs. After 10^{-43} second time becomes reasonably well-defined; when the *horizon* of the Universe (that region in which a causal connection is maintained) reaches a dimension of 10^{-13} centi-meter, the quantum mechanical "size" of a proton, proton-anti-proton production can begin via the reaction (g = graviton):

$$g + g \longrightarrow p^+ + p^-$$

This is but one example of the conversion of gravitational energy into mass. In like fashion, the entire particle-antiparticle density of the early Universe is created.

But what about the small excess fraction of matter (*i.e.*, baryons) that must be present if our current Universe of galaxies is to result after the catastrophic annihilation of the pairs into the primeval fireball of photons? We do not yet know the answer to this question, though at least two possibilities immediately spring to mind: First, the graviton-graviton interaction is probably inherently nonlinear, so that quantum numbers such as the *baryon number* (the number of baryons minus the number of antibaryons) may not be conserved during the turbulent formation period. If this solution is correct, then *lepton number* (referring to electrons, muons, and neutrinos) must also not be conserved, because every excess proton requires an excess electron to balance the charge, for we are fairly certain that the net charge of our Universe is zero. Second, though the baryon number is observed to be conserved to an extraordinary degree of accuracy in the laboratory (approximately one part in 10^{25} per second), the ultrahigh energies involved in the creation of massive pairs, in conjunction with the emphatic arrow of time defined by the rapid universal expansion, may lead to a miniscule violation (about one part in 10^9) of this law in the earliest epoch of our Universe. If the evolution of the Universe can be described by a hot, big-bang model, then it is certain that some such violation of our cherished "laws of physics" must take place. Otherwise, we would not be here now contemplating this problem.

The reader may have noticed that our blithe acceptance of viola-tions of physical laws (at the beginning of the Universe) is strongly reminiscent of the *steady-state theory* of cosmology. In the steady-state theory, neutral matter is continually appearing from nowhere to fill the gaps left by the ever-receding galaxies in the expanding Universe. This phenomenon completely violates the principle of mass-energy conservation, though at a level which is undetectable by direct observations. (To appear more appetizing, a later version of the steady-state cosmology postulates an *ad hoc* negative-energy field, to avoid violating this law.) In our present scheme, it is quite possible that elementary particles are still being created from gravitational energy; but (a) we certainly permit no violation of the law of mass-energy conservation, and (b) the total model is still of the big-bang type.

Very recently, the young mathematician A. Wyler (1971) appears to have succeeded in deriving some of the fundamental properties of elementary particles, using only the symmetry characteristics of special relativistic space-time! If his work is indeed correct, then our model of the beginning of the Universe receives extraordinary support, for this would mean that the fundamental manifestations of matter (*e.g.*, mass, charge, baryon and lepton number) depend upon the geometrical properties of space-time, properties which are grossly distorted at the beginning of a big-bang cosmology (*i.e.*, within about 10^{-43} second of the unmeasurable "beginning"). Hence, we would certainly expect some violation of the known laws of physics when mass is created from gravitational energy. It is not unreasonable to expect slightly more matter to be formed than antimatter, is it?

The Metaphysical Dilemma

Research in quantum cosmology has only just begun. We look forward to the future with great hope and expectation. Perhaps the secrets of our Universe are finally being dragged from the shadows, and the ultimate answers await us just around the corner. However, lest we forget the dangers to which theorizing is prone, let us close by mentioning the metaphysical dilemma.

The questions that we are attempting to answer are metaphysical in nature. If perchance we succeed in creating the theory (that is, a model of reality) that adequately describes the totality of existence, we cannot simply rest satisfied, for reality is unconstrained by our mental constructs. The essence of the metaphysical dilemma is this, that unexpected, that is, unpredictable, phenomena may appear in our Universe at any time. Perhaps we have forgotten to take into account a fifth dimension, which may suddenly intrude itself into our space-time on the day after our so-called "ultimate" theory is completed. We have no control over such a possibility; we must be prepared to accept and attempt to explain any eventuality. Therefore, the job of the scientist (as well as the humanist) is secure, for his task is never-ending.

10. Was There Really a Big Bang?

Geoffrey Burbidge

University of California, San Diego

In any given field of science, activity has its peaks and its valleys. Many view activity as synonymous with progress. If one judges by the number of papers currently appearing and by the interest of astrophysicists, activity in cosmology is at a peak today. To the outsider and even to many inside, it appears at first sight that considerable progress has been made in recent years, so that the outline of the way in which the Universe has evolved is understood, and only the details need to be explained. Is there any justification for this supposition? Those who have read widely in the earlier and more recent cosmological literature, as well as those who rely on J. D. North's (1965) account, are aware that views of cosmology at any epoch are largely determined by the ideas of a few strong individuals rather than by an objective appraisal of the information available. Today a similar situation prevails, and it is further complicated because we are often trying to use classes of objects to test cosmological theories long before we understand the physics of the objects.

In my view, modern developments have been influenced by several factors. These include new observational discoveries combined with an extremely simpleminded approach to them, a deep-seated conviction by many that general relativity is correct in all details, a belief that astronomy has nothing to teach us about fundamental physics, and, last but not least, a strong hostility by both (dis)passionate observers and theoreticians in the astronomical community toward the steady-state theory. Usually this last point is not mentioned in polite society, but it is blatantly obvious in H. Dingle's (1953) remarks while he was President of the Royal Astronomical Society; it is often apparent in private discussions; and it is further manifested by a complete disregard of the theory when the state of cosmology is reviewed (*cf.* I. D. Novikov and Y. B. Zeldovich 1967).

This essay was published in *Nature* 233:36 (1971), and is reproduced with permission of Macmillan Journals Ltd.

The Historical Precedent

What are the major scientific events that make up the case for a big bang? They started, of course, with the work of Albert Einstein in 1917. Incidentally, he, as did his contemporaries, took it for granted that the Universe in the large is unchanging; *i.e.*, he assumed that it is in a steady state. However, it was G. Lemaître (1927, 1931) and, independently, A. Friedmann (1922) who showed that the Einstein model universe was unstable, so that, when perturbed in the direction of expansion, it will go on expanding and will tend to an empty de Sitter universe in the limit. Lemaître's analysis showed that there is an infinity of models satisfying Einstein's equations. Similar solutions are obtained if one uses the scalar-tensor theory of gravitation, which is somewhat different from Einstein's approach but is also consistent with present observations. The most interesting models are those which start at a finite time in a state of infinite density. Scientists therefore inferred that, according to the general theory of relativity, the Universe must be expanding, and that it most likely began in a state of high density. Almost immediately following this major prediction of general relativity, E. P. Hubble (1929) and M. L. Humason (1929) conclusively established the red shift–apparent magnitude relation, which gave observational proof that the Universe is expanding. Thus, by 1930 Einstein's theory had made what many considered to be the greatest prediction in the history of science. For the next twenty years the drive in observational cosmology, almost completely in the hands of Hubble and his colleagues, was to attempt to decide from observation which of the myriad of general relativistic models best represents the Universe. However, there soon developed an apparent paradox associated with the ages of different systems, because it appeared that the ages of various astronomical objects, first of all the Earth, seemed to be greater than the age of the Universe derived from the rate of expansion obtained by Hubble. Two possible paths of escape from this dilemma were the Lemaître model, in which the attractive gravitational force is balanced by a cosmological repulsive force for an arbitrarily long period of extremely slow expansion, and the steady-state cosmology, which requires the continual creation of matter (H. Bondi and T. Gold 1948; F. Hoyle 1948).

The present situation with regard to this age difficulty is that we now realize that the uncertainties associated with the determination of H, the Hubble expansion rate, and those associated with obtaining the ages of the oldest stars are great enough that we no longer have any strong evidence that there is any problem. Most recently it has been stated that $H \sim 50$ kilometers per second per megaparsec, or $H^{-1} \simeq 20 \times 10^9$ years (J. V. Peach 1970; A. R. Sandage 1971), while the age of an old galactic cluster, NGC 188, is given as 10×10^9 years (I. Iben and J. Faulkner 1968; I. Iben and R. Rood 1970; P. Demarque, J. G. Mengel, and M. Aizenman 1971); and the age of the elements obtained from the radioactive isotopes is $\sim 10 \times 10^9$ years, with the uncertainties depending on what one assumes about the sequence of events leading to the isolation of uranium and thorium in the Solar System (W. A. Fowler 1968).

The prediction that the Universe is expanding, followed by the discovery of the expansion, was the first major success of cosmology based on general relativity. The second came from the comparatively elementary discussion of the physics of the early stage in a dense expanding universe, the primeval fireball by G. Gamow (1953) and his colleagues (R. A. Alpher and R. Herman 1949), and the discovery in 1965, by A. Penzias and R. Wilson, of background radiation, which may well be relict radiation from this fireball. These two predictions, followed by two major discoveries that, at first sight, appeared to confirm the theory, provide the basic scientific case for evolving models of the Universe.

If we wish to account for the expansion without invoking an initially dense phase, we are left with only the steady-state model, in which matter is created, or with an oscillating model. The steady-state alternative seemed attractive to some, especially, as F. Hoyle (1948) originally pointed out, "when taken in conjunction with aesthetic objections to the creation of the universe in the remote past. For it is against the spirit of scientific inquiry to regard observable effects as arising from causes unknown (and unknowable) to science, and this in principle is what creation in the past implies." There is apparently a great gulf between those who find creation in the past acceptable but continuous creation unacceptable and those who have no such strong feelings. It is my belief that much of the conflict in cosmology in modern times has arisen because of this division, which has little to do with science. More to the point, we must critically evaluate the detailed evidence that has accrued in the last twenty years concerning the question of whether or not we have definitely established that the Universe is evolving.

Determination of q_o

In 1956 M. L. Humason, N. U. Mayall, and A. R. Sandage gave a detailed account of the status of the Hubble relation. At that time the largest red shifts that had been measured were $z \simeq 0.2$. In the fifteen years that have elapsed since then, Sandage and his collaborators have worked intensively on the problem of refining the measurements of H and q_o. Many progress reports on this work have been given by Sandage (1958, 1961, 1966, 1968, 1970); see also J. V. Peach (1970). As far as the value of H is concerned, there is some reason to believe that we are beginning to converge toward a realistic value some ten times smaller than that obtained by Hubble in 1936. (See the account by Sandage of the succession of corrections that have been applied.) As far as the determination of the deceleration parameter q_o is concerned, the position is not so good. There are two basic reasons for this. First, since Humason pushed out to red shifts of 0.2, very few clusters with greater red shifts have been discovered. The most recent plot given by Sandage (1970) contains 42 first-ranked galaxies in clusters, and only three of these, with $z = 0.29$, 0.36, and 0.46, are in excess of 0.2. The three clusters were observed by W. A. Baum (1957, 1961) more than ten years ago, and a detailed discussion of his data has never been given. Many people believe that the radio astronomer's ability to detect more distant objects will lead, or has led, to an improvement of the situation. However, the only major achievement so far of radio astronomy in this connection was the identification, in 1960, of 3C 295 with a galaxy with a red shift of $z = 0.46$, measured by R. Minkowski. This source lies in a cluster, and Baum obtained photometry that, combined with the red shift, gave this last point on the Hubble diagram.

About 100 galaxies associated with radio sources and 200 quasi-stellar objects, most of which are radio sources, have been optically identified, and their red shifts have been obtained. None of these, with the exception of the cluster around 3C 295, has given us any information about the classical Hubble diagram beyond $z \sim 0.2$. This is because the radio sources are associated with galaxies with red shifts less than 0.2, or they are associated with galaxies that obviously cannot be treated as "standard candles," or they have been identified with quasi-stellar objects that show a very large dispersion in the red shift–apparent magnitude diagram. Using the radio properties as a guide, we single out a certain class of objects, but it is not clear at all that they can be used for cosmological investigations of the classical kind. If we make a table of optical types of radio galaxies as a function of increasing red shift (G. R. Burbidge 1970), it is seen that, at small red shifts, practically all types of galaxy are being identified. As one goes to intermediate red shifts the galaxies are mostly giant ellipticals, while at the largest red shifts, between 0.2 and 0.3, the galaxies with published red shifts have nearly all been classified as N-type, meaning that a large part of the optical energy that they are emitting is due to nonthermal processes and hot gas rather than stars. Thus they cannot be used as standard candles, as the data are available at present.

So far it has not been possible to make optical identifications and obtain red shifts of radio sources that are fainter than about 18^m* in the case of nonstellar objects (galaxies) or about $19\overset{m}{.}5$ in the case of stellar objects. These are limits set by optical astronomy. They are set, first, by the fact that the 48-inch Palomar Sky Survey only goes down to $20\overset{m}{.}0$ in the red and $21\overset{m}{.}1$ in the blue, and objects that are fainter than this cannot be identified unless the large reflectors that typically have very small fields of less than 20 arc-minutes are used. Second, the difficulties associated with studying very faint objects are very great, in part simply because the night-sky background against which one is observing amounts to about 21^m per square arc-second, so that it is increasingly difficult to detect photons against the natural and manmade noise, even with the use of image-tube techniques; and partly because only a very limited fraction of observing time on the large telescopes is presently devoted to this work.

Even if many of these problems are overcome and it eventually becomes possible to measure the red shifts and to do photometry of galaxies down to 21^m to 23^m, we will still have a problem in finding suitable objects to observe, since there is no prospect of doing surveys to this level. Sandage (1970) and his colleagues (J. Kristian and A. R. Sandage 1970) are now attempting to make identifications by using very accurate radio source positions for which no optical object can be seen at the level of the 48-inch Palomar Sky Survey. By using special photographic plates and the 200-inch telescope, they have been able to go as faint as 22^m to 23^m and find optical objects in some of these positions. Even when such objects are

---

*Superscript *m* denotes *magnitude*, a measure of brightness; the larger the magnitude, the fainter is the object.

found, measuring red shifts and energy distributions will be very time-consuming, and it is not obvious that it will succeed. Those objects that turn out to be quasi-stellar objects will be useless for this type of investigation. We can use the red shifts of those that turn out to be galaxies, but even then it will be necessary to observe other galaxies that are associated with them to obtain energy distributions that are likely to be sufficiently normal so that they can be used to derive apparent *bolometric magnitudes* (which measure their total luminosity over all wavelengths).

A second problem concerns the corrections to the observed energy distribution, based on the view that we are looking at standard galaxies, if they exist. These, at large red shifts, are much younger than the galaxies comparatively close to us, from the viewpoint of stellar evolution. To make such corrections it is necessary to know what types of stars give rise to the bulk of the radiation from elliptical galaxies. A limited amount of work has now been done on the nuclear regions of nearby spirals and on one or two ellipticals in the Virgo cluster. This has led Sandage to conclude that the evolutionary corrections may amount to no more than $0^{m}.03$ per 10^9 years, a value considerably smaller than it was estimated to be some years ago. B. Tinsley has argued that theoretical models of galaxies that show a greater degree of evolution, so that the color correction may be considerably larger, are also compatible with the observations. Most observers are doubtful of her conclusions, but it is far from clear that the present results are conclusive.

At a red shift of $z = 0.2$, the difference in bolometric magnitude of the standard galaxies, m_{bol}, to be expected between $q_o = 0$ and $q_o = +1$, is only $0^{m}.2$, while at $z \simeq 0.5$, the difference in m_{bol} between $q_o = +1$ and $q_o = -1$ (steady-state) is $0^{m}.9$.

Since the three clusters with red shifts greater than 0.2 appear to lie closest to the $q_o = +1$ relation, it has been stated for the last ten years, with different degrees of emphasis, that $q_o \simeq +1$, although at the same time it also has sometimes been implied that q_o is not really known at all. In terms of hard data, this latter statement is true. The facts that Baum's measurements have never been published in detail and have never been repeated, that we have such a small amount of data overall, and that the local anisotropy in the velocity field may also affect the determination of q_o all suggest that the numerical value of q_o is only known to lie between wide limits, so that both the exploding models and steady-state models are compatible with the observations made so far.

Stebbins-Whitford Effect

In 1948, J. Stebbins and A. E. Whitford published the first results of their photoelectric measurements of the colors of faint clusters of galaxies. They concluded in this preliminary investigation that after they had put in the usual corrections for the effect of the red shift there appeared to be an excess reddening which increased with increasing red shift. This meant that the distribution of energy radiated by stars in a galaxy is a function of epoch, so that an evolutionary effect is present. In the early 1950's, this result was used as observational evidence against the steady-state cosmology, and there was some discussion in the literature about its significance (H. Bondi, T. Gold, and D. Sciama 1954; A. E. Whitford 1954). However, the result was only preliminary and by 1956 Whitford no longer believed that it was correct. It took until 1968 for J. B. Oke and Sandage, in a properly published investigation, to put this result to rest. Unfortunately, for many years after the realization by many observers that the result was spurious, theoreticians were using it as direct evidence against the steady-state theory.

Counts of Radio Sources

If any class of extragalactic objects is distributed uniformly in space, counts to successively increasing metric distance, and hence to fainter apparent brightnesses, will give numbers that are proportional to the volumes. The volumes in Riemannian space depart from the volumes calculated on the Euclidean approximation by factors that are determined by the space curvature. Hubble (1936) first attempted to count optical galaxies to apply this test but he abandoned the attempt, and more recently Sandage (1961) has shown that, unless we go to much fainter galaxies than can be observed from the Earth, attempts to distinguish between different cosmological models will not succeed.

However, the method has been applied extensively to extragalactic radio sources by several groups who have counted sources at a number of frequencies down to different minimum flux* levels. The

---

*When a radio receiver "looks" at a given area (steradians) on the sky, the energy that it detects each second (watts) per unit area (square meters) of the receiver per unit interval of frequency (Hertz) is the *flux density* S. One flux unit = 1 f.u. = 10^{-26} watt meter^{-2} steradian^{-1} Hertz^{-1}.

early work by M. Ryle (1955) and his colleagues (M. Ryle and P. A. G. Scheuer 1955) gave a log N – log S (number *versus* flux) slope for the brighter sources of -3, far steeper than the expected Euclidean value of -1.5. (If N is the number of sources per steradian brighter than S, and R is the distance to a source, then $N \propto V \propto R^3$, $S \propto R^{-2}$, so that in Euclidean space $N \propto S^{-3/2}$. For distant sources in Riemannian space, $N \propto S^{-3/2 + \mu(S)}$ where $\mu(S) \geq 0$ is a function depending on the maximum source distance and the cosmological model.) Quite early it became clear that the first log N–log S slope obtained in Cambridge was far too steep, and other surveys (B. Y. Mills 1959; G. A. Day *et al.* 1966; M. Ryle and P. F. Scott 1961) all led to the conclusion that at the bright end of the curve the slope is steeper than -1.5 and has a value somewhere between -1.75 and -1.85. Following Ryle and his associates it has commonly been argued that this steep slope can only be explained as being due to an excess of sources at great distances and that this requires evolution of the source numbers with epoch. This is then evidence against the simple steady-state theory. Observations to fainter flux levels show that the log N–log S plot gradually flattens to a slope of only ~ -0.8 at $S \simeq 0.01$ flux units. All these results have been obtained by making surveys at comparatively low frequencies, typical values being 408 MHz and 178 MHz. The work that has been done at higher radio frequencies, 2700 MHz, leads to a log N–log S slope of -1.4 ± 0.1 (A. T. Shimmins, J. Bolton, and J. V. Wall 1968). Were this the only radio survey no one would have argued that significant departures from a homogeneous universe exist. However, it has been argued that because of the wide dispersion in spectral indices of the radio sources and their variation with flux density, the slope of the log N–log S curve should decrease with decreasing frequency of observation. The lower frequency survey containing a more homogeneous source sample would then be the one most representative of the gross features of the Universe. This situation naturally leads to the following unsolved problem: at what observing frequency (radio, microwave, infrared, optical, X-ray) is an unbiased sample most likely to be selected? It would really be preferable to integrate the flux over a suitably large bandwidth and then construct a log N–log S integrated curve.

How can we interpret the total slope of the log N–log S curve for sources measured at low frequencies? Ryle (1968) and his associates (M. Longair 1966) have argued that these results indicate that the Universe must be evolving and that there was an excess of sources in an epoch defined by $z \gtrsim 2$, but that at even earlier epochs the source density was much lower again. Two alternative proposals are either that the slope for bright sources does not depart significantly from the Euclidean value or that its steepness is due to a statistical fluctuation in the number of very bright sources, requiring a deficit of from 5 to 7 sources per steradian. Hoyle (1969) has argued that the remainder of the curve is directly compatible with the prediction of the steady-state model. Ryle and his associates disagree with this and argue that the whole curve can be explained only by using Friedmann models with source evolution. The arguments depend on the details of the radio luminosity function and the exact slope of the log N–log S curve for faint sources.

We have optical identifications and red shifts for only a very small fraction of the extragalactic radio sources. However, if we restrict ourselves to the 3C (revised) catalog (A. S. Bennett 1962) containing about 250 of the brightest radio sources ($S \geq 9$ f.u.) a total of about 40 percent of the sources, including about 40 quasi-stellar objects and 68 galaxies, have been identified and red shifts have been obtained. The identification of the quasi-stellar objects appears to be complete; so the remainder of the sources are thought to be associated with galaxies for which identifications have been made but for which no red shifts have been obtained or, in a minority of cases, for which there are no optical objects visible on the Palomar Sky Atlas at the radio positions. Plots of radio flux S against red shift have been made both for the quasi-stellar objects and the galaxies (F. Hoyle and G. R. Burbidge 1966, 1970; M. Longair and P. A. G. Scheuer 1967; J. G. Bolton 1969), and in each case it has been found that there is no correlation between the apparent flux and the red shift. Thus it appears that for the identified sources the log N–log S curves are due to a luminosity effect, not a distance-volume effect. The key to the interpretation of the log N–log S curve for the 3C sources lies in the remaining radio galaxies. If they turn out to have red shifts systematically larger than those already measured, as has been generally assumed because they are optically fainter, then their log N–log S curve is of cosmological significance. However, if they do not have systematically greater red shifts it must be concluded that the log N–log S curve is dominated by the effects of the radio luminosity function in a comparatively local region. The quasi-stellar objects do have large red shifts. If these are of cosmological origin then their log N–log S curve and the luminosity-volume test (M. Schmidt 1968) do demonstrate a powerful evolutionary effect. However, if the red shifts are not of cosmological origin, these analyses give us no evidence of cosmological importance.

A more detailed discussion of the analyses of the counts of sources has been given by K. Brecher, G. Burbidge, and P. Strittmatter. The limitations set by the source counts on various cosmological models are summarized in Table 10.1, which is taken from their paper. It can be seen from this table that, contrary to popular opinion, the counts of radio sources do not rule out a steady-state universe.

The Quasi-Stellar Objects

The quasi-stellar objects are the only extragalactic objects so far discovered that have very large red shifts. Therefore they are of the greatest importance for cosmology provided that their red shifts are largely a result of the expansion of the Universe and do not have any other explanation. If the red shifts are of cosmological origin, then the luminosity-volume test previously mentioned for these objects in the 3CR catalog, or a similar analysis also carried out by Schmidt (1970) for a sample of quasi-stellar objects that are not powerful radio sources, leads to the conclusion that they are not distributed uniformly in space, but that the space density is very strongly dependent on z, and hence on epoch. This means that they provide powerful evidence against a simple steady-state universe, and if one wishes to preserve the steady-state model it must be argued that it is a fluctuating steady-state universe in which evolution can occur within an epoch corresponding to red shifts up to ~ 2. In any case the cut-off of red shifts beyond 2 is of great cosmological significance, and some have attempted to relate this to the appropriate epoch in a universe of Lemaître type (V. Petrosian, E. E. Salpeter, and P. Szekeres 1967; I. S. Shklovsky 1967).

However, none of these arguments has any validity unless it can be demonstrated conclusively that the red shifts of the quasi-stellar objects are of cosmological origin, and this has simply not been done. There are a number of arguments that can be adduced for and against the cosmological hypothesis, but there is no certain answer. Until this is reached, use of quasi-stellar objects for cosmological investigation is premature.

TABLE 10.1. Limitations Set by Source Counts on Various Cosmological Models

Source	Strict Steady-State	Fluctuating Steady-State	Friedmann	Friedmann with Source Evolution
Quasi-stellar objects				
$z \rightarrow$ Distance	Excluded	Allowed[a]	Excluded	Allowed[b]
$z \nrightarrow$ Distance	Allowed	Allowed	Allowed	Allowed
Radio galaxies				
z known				
$z \rightarrow$ Distance	Allowed	Allowed	Allowed	Allowed
z unknown				
$m_v \rightarrow$ Distance	Excluded	Allowed[c]	Excluded	Allowed[c]
z unknown				
$m_v \nrightarrow$ Distance	Excluded[d]	Allowed	Excluded[d]	Allowed

[a] This case is allowed only if fluctuation to $z \sim 2$ is permitted.
[b] The observed background limits such evolution to $2 < z(QSO, \text{max}) < 4$.
[c] The observed background limits such evolution to $z(RG, \text{max}) < 0.6$.
[d] This case would require a special, rather implausible, luminosity function to avoid exclusion.

The Microwave Background Radiation

We earlier pointed out that the discovery of microwave background radiation, following the prediction that primeval fireball radiation generated in a big bang would be present, was one of the strongest pieces of evidence for such a beginning. Is the information that we now have on the microwave background still compatible with this interpretation? Observations made at wavelengths between a few millimeters and about 20 centimeters fit quite well on a blackbody curve with a temperature of about 2.7°K. The energy density of the radiation is then about 4×10^{-13} erg per cubic centimeter. If the blackbody form is confirmed, this will be the strongest evidence in favor of an evolving universe. However, it is not yet certain whether or not the spectrum is really of black body form, since the wavelength range directly observed, and not in dispute, is the Rayleigh-Jeans part of the curve where $I(\nu) \propto \nu^2$.

The observations that have been made either directly or indirectly close to, and just beyond, the peak of the blackbody curve are in conflict unless it is argued that some very strong line radiations are present. The indirect observations of the radiation field (in the galactic plane) that is exciting the CN, CH, and CH$^+$ interstellar molecules suggest that the temperature is less than $\sim 8°K$ at 9.36 millimeters, less than 5°K at 0.56 millimeter, and less than $\sim 4°K$ at 1.3 millimeters, and it appears to be about 2.8°K at 2.64 millimeters. However, direct observation from rockets and a balloon above the Earth's atmosphere at wavelengths in the range 0.4 to 1.3 millimeters gives a flux which corresponds to a blackbody temperature of about 6°K (K. Shivanandan, J. R. Houck, and M. Harwit 1968; J. R. Houck and M. Harwit 1969; D. Muelhner and R. Weiss 1970). It is possible that the radiation detected by these observers is concentrated in a line which could be produced in a geocorona, in the Galaxy, or in the extragalactic arena. The interpretation of the results is unclear at present.

If it turns out that the microwave radiation field departs strongly from the blackbody form, then either extra components are present, or the radiation has a quite different origin. In both cases the answer must be that the excess radiation, or the total flux, ultimately arises in discrete sources. A number of investigations have been made in which these possibilities have been explored. The sources are most likely to be the nuclei of galaxies that radiate largely at infrared and microwave frequencies.

As was mentioned earlier, the discovery of the microwave background radiation by Penzias and Wilson in 1965 undoubtedly has had a great impact on cosmology. It was this discovery that led to the large number of theoretical investigations concerned with the early history of the Universe. What should be made clear, however, is that the existence of this radiation is very powerful evidence for an initially dense state, provided that the radiation has a blackbody spectrum as is predicted in the simple theory. If it turns out that the spectrum departs strongly from the blackbody shape, many attempts will undoubtedly be made to explain the result in terms of a primordial radiation field plus large contributions from discrete sources, or even a more complex big bang. However, this must not obscure the important point that the absence of a blackbody form would mean that the prediction has not been fulfilled, and the strong and direct evidence for an initial dense state will have disappeared.

This concludes our discussion of direct observational evidence bearing on whether or not the Universe is evolving and began in a dense state. It would appear that if one attempts to evaluate this evidence objectively there is still no really conclusive evidence in favor of such a universe.

Nucleosynthesis and the Origin of Galaxies

Another approach can be taken in looking for evidence that bears on the question of whether or not the Universe began in a dense state. The bulk of the mass that we see is condensed in the form of stars and galaxies with a wide range of masses and angular momenta. This matter is not distributed uniformly in space but is condensed into dense aggregates. Moreover it has a complicated chemical composition. The major problem of cosmogony is therefore to understand the origin of the elements and the formation of discrete objects—galaxies and other compact massive objects.

Does the existence of these features of the Universe suggest that their origin is closely connected with a big bang?

Let us consider first the problem of nucleosynthesis. Early work on this problem made little progress because knowledge of nuclear physics was still comparatively primitive. It was G. Gamow (1946, 1948), R. A. Alpher (1948), and Alpher and R. C. Herman (1948, 1950) who first made a serious attempt to explain the synthesis of the heavier elements from an initial dense cloud of baryons, leptons, and radiation in a hot big bang. It is well known that this attempt failed due to the difficulty associated with building beyond mass 4. In the following decade a serious attempt was made to understand the relative abundances of all the isotopes on the assumption that they have nearly all been synthesized in the interiors of stars (E. M. Burbidge, G. R. Burbidge, W. A. Fowler, and F. Hoyle 1957). This theory has been, in large part, highly successful and is generally accepted. It was originally thought that the fact that heavy elements are synthesized from lighter ones in stars meant that there is a slow buildup of heavy elements as a galaxy grows older, and that the relative abundance would be correlated with the ages of the stars. It is now clear that while this effect is operating, a large amount of nucleosynthesis takes place early in the life of a galaxy and the processes going on steadily throughout its life are in some ways of lesser importance.

From the viewpoint of cosmology the most important result is that very detailed calculations made by R. Wagoner, W. A. Fowler, and F. Hoyle (1967) have demonstrated conclusively that the bulk of the elements cannot have been made in a big bang. The only important element that, from the viewpoint of nuclear physics, could have been synthesized in a big bang is helium. Thus, soon after the discovery of the microwave background radiation a serious attempt was made to argue that the helium detected in stars and gaseous nebulae in our own Galaxy and in other galaxies is primordial helium made in a big bang. The argument appeared plausible, since, on the one hand, it was suggested that the helium/hydrogen ratio is the same in different celestial objects (stars, gaseous nebulae in our own Galaxy, and a few gaseous nebulae in one or two nearby galaxies) and is 25 to 30 percent by mass while, on the other hand, the fractional amount of helium relative to hydrogen that is made in a big bang is close to about 27 percent. However, this simple argument broke down as soon as detailed studies of different aspects of the problem were made. First of all, it was pointed out that there appear to be stars that contain very little helium, so that perhaps there is no universal helium abundance. While this argument is still open, detailed studies of big-bang models led to the conclusion that there are ways of making very little helium in such an initial process (S. W. Hawking and R. J. Tayler 1966; K. Thorne 1967; R. Dicke 1968). Fowler (1970) has recently emphasized that the helium/hydrogen ratio is determined completely by the choice of baryon and lepton numbers in a big bang, something which has nothing to do with the general theory of relativity but is tied only to initial conditions that lie outside the known laws of physics. Thus, while the helium abundance in the Universe might be related to the occurrence of a big bang, there is simply no direct evidence that it is, and there are a number of alternative ways of understanding the large helium abundance found in some stars and galaxies. For example, it appears that helium could have been synthesized in massive objects evolving in the nuclear regions of galaxies.

Finally we come to the vexed question of the origin of galaxies. As the big-bang bandwagon has gained momentum, an increasing number of investigations have been carried out attempting to explain the condensation of dense objects from an initial cloud of matter and radiation which is expanding. It has been known for many years that this is very difficult to understand (E. Lifshitz 1946; W. B. Bonnor 1956; E. Lifshitz and I. M. Khalatnikov 1963), and the investigations have now reached the point where it is generally accepted that the existence of dense objects cannot be understood unless very large density fluctuations in a highly turbulent medium, or otherwise, are invoked in the first place (E. R. Harrison 1970; L. M. Ozernoy and A. D. Chernin 1967, 1968). There is again no physical understanding of the situation; it is a condition which is put in, in a hypothetical state, to explain a major property of the Universe. Thus these "theories" amount to nothing more than the statement that protogalaxies have a cosmological origin, and their origin cannot be understood any better than can the original baryons and leptons in an evolving universe. At this level, therefore, these theories are in no better shape than the apparently much more radical views of V. A. Ambartsumian (1958, 1965), or F. Hoyle and J. V. Narlikar (1966), who believe that the origin of galaxies is tied closely to the properties of the dense active nuclei and that perhaps the initial states of galaxies were high density states. Hoyle and Narlikar have even dared to suggest that perhaps "new" physics is involved. The debate largely reduces to two viewpoints: an apparent belief by the majority that creation in the distant past is acceptable, but that creation at recent epochs is unthinkable, and an opposing opinion that there is little to choose between the two alternatives, and that only with much more original work can we hope to resolve the issue. What is clear is that the evolving universe concept gains no support from attempts to understand either the origin of the elements or the origin of galaxies.

Conclusion

Was there really a big bang? I believe that the answer clearly must be that we do not know. If we are ever to find an answer much more effort must be devoted to cosmological tests, with a much more open-minded approach, and much more original thinking must be done to attempt to explain the large amount of observational material, and not just that material which can be used in a narrow sense to fit preconceived ideas. Probably the best argument in favor of a beginning is perhaps the general result that the ages of many stars in our Galaxy are approximately equal to H^{-1}. Probably the strongest argument against a big bang is that when we come to the Universe in total and the large number of complex condensed objects in it, the theory is able to explain so little.

Bibliographical Commentary

An Introduction to the Emerging Universe

Semipopular articles on modern astronomy often appear in *Scientific American*, *Sky and Telescope*, and the daily press. Of particular interest are *The New Astronomy*, edited by the editors of *Scientific American* (New York: Simon & Schuster, 1955), and *Frontiers In Astronomy*—readings from *Scientific American* with introductions by O. Gingerich (San Francisco: W. H. Freeman and Co., 1970). Good elementary textbooks on the subject are: *Exploration of the Universe* by George Abell (2d ed., New York: Holt, Rinehart & Winston, 1969), *Astronomy* by R. H. Baker and L. W. Fredrick (9th ed., New York: Van Nostrand Reinhold, 1971), and *Elementary Astronomy* by O. Struve, B. Lynds, and H. Pillans (New York: Oxford University Press, 1959). For another point of view, see *Galaxies, Nuclei and Quasars* by Fred Hoyle (New York: Harper & Row, 1965), and for a picture of the current commotion in astronomy, *The Violent Universe* by Nigel Calder (New York: Viking Press, 1970). *A History of Astronomy from Thales to Kepler* by J. L. E. Dreyer (New York: Dover Publications, 1953) is an excellent account which emphasizes the long, slow development of astronomy.

The quotation of John Herschel is taken from an account of the early history of the Royal Astronomical Society by G. J. Whitrow, *Quarterly Journal of the Royal Astronomical Society* 11: 89 (1970). The concluding quotation is from Bertrand Russell's essay "The Expanding Mental Universe," *The ABC of Relativity* (London: George Allen & Unwin, 1925).

1. Why Does the Sun Shine?

A great classic monograph in the field of stellar structure is A. S. Eddington's *The Internal Constitution of the Stars* (Cambridge: The

University Press, 1926; reprinted New York: Dover Publications, 1959). Though originally published in 1926, its lucid exposition of fundamentals and delightful technical style make it worth reading today. M. Schwarzschild's *Structure and Evolution of the Stars* (Princeton, N.J.: Princeton University Press, 1958; reprinted New York: Dover Publications, 1966) gives a clear mathematical discussion of the methods used to analyze the interiors of stars. See also L. H. Aller's *Atoms, Stars and Nebulae* (Cambridge, Mass.: Harvard University Press, 1971). *The Evolution of Stars*, edited by T. Page and L. W. Page (New York: Macmillan, 1968), is an excellent comprehensive collection of semitechnical articles from *Sky and Telescope* on recent developments in this subject, and *Stellar Evolution* (New York: Pergamon, 1967) by A. J. Meadows is a nonmathematical discussion of the recent state of the art. Freeman Dyson's recent semipopular article may be found in *Scientific American* 224: 50 (September 1971).

2. *Planetary Systems*

A brief, but quite comprehensive, survey of theories relating to the origin of our Solar System has been given by I. P. Williams and A. W. Cremin, *Quarterly Journal of the Royal Astronomical Society* 9: 40 (1968). See also the interesting ideas of H. Alfvén and C. Arrhenius, *Astrophysics and Space Science* 8: 338 (1970), and B. J. Levin and V. S. Safronov, *Theory of Probability and Its Applications* 5: 220 (1960). H. P. Berlage's *The Origin of the Solar System* (New York: Pergamon Press, 1968) is a semipopular discussion which emphasizes the theory of the turbulent primordial nebula.

Other useful references are G. P. Kuiper's chapter in *Astrophysics*, ed. J. A. Hynek (New York: McGraw-Hill, 1951), and O. Struve's *Stellar Evolution* (Princeton, N.J.: Princeton University Press, 1950).

Intermediate and advanced works, directly related to the nonstandard ideas presented in this essay, are:

S. S. Kumar, *Astrophysical Journal* 137: 1121 (1963).

———, *Zeitschrift für Astrophysik* 58: 248 (1964).

———, *Icarus* 6: 136 (1967).

———, *Annals of the New York Academy of Sciences* 163: 94 (1969).

P. J. Shelus, *Celestial Mechanics* (paper submitted in 1971).

P. J. Shelus and S. S. Kumar, *Astronomical Journal* 75: 315 (1970).

C. Worley, a chapter in *Low Luminosity Stars*, ed. S. S. Kumar (New York: Gordon & Breach, 1969).

Other references cited in the essay are:

S. S. Huang, *Astrophysical Journal* 150: 229 (1967).

R. P. Kraft, *Astrophysical Journal* 150: 551 (1967).

3. *The Crab Nebula and Pulsar*

The early Chinese records and the evidence for the connection of the Crab nebula with the 1054 A.D. event are discussed by J. J. L. Duyvendak, *Publications of the Astronomical Society of the Pacific* 54: 91 (1942).

The classic work on the Crab nebula, as it was understood before the discovery of pulsars, is L. Woltjer, *Bulletin of the Astronomical Institute of the Netherlands* 14: 39 (1958). Much of the same material is covered at a less technical level by N. U. Mayall, *Science* 137: 91 (1962).

Supernovae and their remnants in general are discussed in a textbook by I. S. Shklovskii, *Supernovae* (London: John Wiley & Sons, 1968), and in P. J. Brancazio and A. G. W. Cameron (eds.), *Supernovae and Their Remnants* (New York: Gordon & Breach, 1969), which is the report of a conference held in November 1967. The former suffers from a large number of mistranslations, some of which are amusing and some merely confusing. The latter volume contains a very important series of photographs of the Crab nebula which illustrate changes in its wisp structure over the past twenty years. Both books were somewhat out of date even at the time they were published.

Since the discovery of pulsars, two conferences have been held on the subject of the Crab nebula. Their proceedings are published in *Publications of the Astronomical Society of the Pacific* 82: 375 ff. (1970) (the Flagstaff conference, held in June 1969), and in R. D. Davies and F. G. Smith (eds.), *IAU Symposium No. 46—The Crab Nebula* (Dordrecht, Holland: D. Reidel, 1971) (the Manchester conference, held in August 1970). Both volumes contain reviews of observations and theory (with fairly complete references to the literature), as well as reports of research in progress at the time of the conferences. The report of a symposium on pulsars and supernova remnants held at Rome in December 1969 is, unfortunately, still "in press."

4. Chemistry between the Stars

Popular expositions of the current state of astrochemistry in the interstellar medium may be found in the following scientific articles: L. E. Snyder and D. Buhl, "Molecules in the Interstellar Medium," *Sky and Telescope* 40: 267 (November 1970) and 345 (December 1970); D. Buhl and L. E. Snyder, "From Radio Astronomy towards Astrochemistry," *Technology Review* 73: (April 1971); D. Buhl and C. Ponnamperuma, "Interstellar Molecules and the Origin of Life," to be published soon in *Space Life Sciences*.

Background material fundamental to this essay may be found in these books: P. W. Merrill, *Space Chemistry* (Ann Arbor: University of Michigan Press, 1963); A. G. Pacholczyk, *Radio Astrophysics* (San Francisco: W. H. Freeman, 1970); G. Herzberg, *Molecular Spectra and Molecular Structure* (New York: G. Van Nostrand, Vol. I, 1950; Vol. II, 1964; Vol. III, 1966); C. H. Townes and A. L. Schawlow, *Microwave Spectroscopy* (New York: McGraw-Hill, 1955); and B. T. Lynds (ed.), *Dark Nebulae, Globules, and Protostars* (Tucson: University of Arizona Press, 1971).

A fascinating work of science fiction related to the problems considered here is: F. Hoyle, *The Black Cloud* (New York: Harper & Row, 1958).

In addition, the following literature is referred to in this essay:

1. A. C. Cheung, D. M. Rank, C. H. Townes, D. D. Thornton, and W. J. Welch, *Physical Review Letters* 21:1701 (1968).
2. ———, *Nature* 221:626 (1969).
3. L. E. Snyder, D. Buhl, B. Zuckerman, and P. Palmer, *Physical Review Letters* 22:679 (1969).
4. P. Palmer, B. Zuckerman, D. Buhl, and L. E. Snyder, *Astrophysical Journal Letters* 156: L147 (1969).
5. L. E. Snyder, *Spectroscopy, Vol. 1, Biennial Review of Chemistry* (MTP International Review of Science), ed. D. A. Ramsay, Chapter 9 (1972).
6. D. Buhl and L. E. Snyder, *Nature* 228:267 (1970).
7. D. E. Milligan and M. E. Jacox, *Journal of Chemical Physics* 39:712 (1963).

5. Life in the Universe

The great modern introduction to biochemistry is J. D. Watson's *Molecular Biology of the Gene* (New York: Benjamin, 1965), which can be understood if one knows elementary chemistry. *Time's Arrow and Evolution* by H. F. Blum (New York: Harper & Row, 1962) is an extended and stimulating essay on the origin and properties of living systems. The astrophysical requirements for human life are described in a quantitative way by S. H. Dole in *Habitable Planets for Man* (New York: Blaisdell, 1964). *Intelligent Life in the Universe* by I. S. Shklovskii and C. Sagan (New York: Dell, 1966) is both a good introduction to modern astronomy and a thorough discussion of the problems of discovering extraterrestrial life. A number of interesting, not very mathematical, articles appear in *Interstellar Communication*, ed. A. G. W. Cameron (New York: Benjamin, 1963).

Excellent popular accounts of chemical evolution and living organisms appear in *Scientific American* from time to time. For an excellent overview of the ground covered in this essay, we highly recommend *The Scientific Endeavor*, ed. D. W. Bronk and F. Seitz (New York: Rockefeller Institute Press, 1965); Robert Jastrow's *Red Giants and White Dwarfs* (New York: Harper & Row, 1967); and *The Origin of Life* by A. I. Oparin (New York: Dover Publications, 1953).

6. Galaxies: Landmarks of the Universe

The *Atlas of Peculiar Galaxies* by H. Arp (Pasadena: CalTech Bookstore, 1966; also reprinted in *Astrophysical Journal Supplement* 14: 1, 1966) illustrates 338 anomalous, often interacting, galaxies. A discussion of the connected pair of galaxies with greatly different red shifts in Figure 22 is also given by H. Arp in *Astrophysical Letters* 7: 221 (1971). For a beautiful atlas of primarily normal galaxies, which also describes the Hubble classification scheme, see the *Hubble Atlas of Galaxies* by A. R. Sandage (Washington, D.C.: Carnegie Institute, 1961).

A reprint of E. Hubble's Silliman Lectures first published in 1936 is available in *The Realm of the Nebulae* (New York: Dover Publications, 1958). Although the quantitative data have been expanded and improved, this is still the definitive introduction to extragalactic astronomy. A variety of recent reviews and papers may be found in *External Galaxies and Quasi-Stellar Objects, International Astronomical Union Symposium No. 44*, edited by D. S. Evans (Dordrecht, Holland: D. Reidel, 1971). For an excellent recent review of the development of the subject, there is J. D. Fernie's "The Historical

Quest for the Nature of the Spiral Nebulae" in *Publications of the Astronomical Society of the Pacific* 82: 1189 (1970). A basic paper on the determination of colors and magnitudes of galaxies by E. Holmberg is in No. 136 of Series II of *Meddelande från Lunds Astronomiska Observatorium* (1958). For recent summaries of our understanding of the stellar populations, masses, hydrogen content, radii, and luminosities of galaxies, see the article by I. R. King in *Publications of the Astronomical Society of the Pacific* 83: 377 (August 1971), and that by M. S. Roberts in the *Astronomical Journal* 74: 859 (1969).

7. What Olbers Might Have Said

H. Bondi, *Cosmology* (Cambridge: Cambridge University Press, 1952).

W. B. Bonnor, "On Olbers' Paradox," *Monthly Notices R.A.S.* 128: 33–47 (1964).

C. V. L. Charlier, *Arkiv foer Matematik, Astronomi och Fysik* 16, no. 22: 1–34 (1922).

F. P. Dickson, *The Bowl of Night* (Eindhoven, Holland: Philips Technical Library, 1968).

E. R. Harrison, "Olbers Paradox and the Background Radiation Density in an Isotropic Homogeneous Universe," *Monthly Notices R.A.S.* 131: 1–12 (1965).

F. Hoyle, "A New Model for the Expanding Universe," *Monthly Notices R.A.S.* 108: 372–82 (1948).

W. H. McCrea, "A Philosophy for Big-Bang Cosmology," *Nature* 228: 21–24 (1970).

J. D. North, *The Measure of the Universe* (Oxford: Clarendon Press, 1965), p. 18.

D. T. Pegg, "Night Sky Darkness in the Eddington-Lemaître Universe," *Monthly Notices R.A.S.* 154: 321–27 (1971).

F. E. Roach and L. L. Smith, "An Observational Search for Cosmic Light," *Geophys. J.* (G B) 15: 227–39 (1968).

D. W. Sciama, *The Unity of the Universe* (London: Faber & Faber, 1959).

G. de Vaucouleurs, "The Case for a Hierarchical Cosmology," *Science* 167: 1203–12 (1970).

E. T. Whittaker, *A History of the Theories of Aether and Electricity: The Classical Theories* (London: Nelson, 1951), chap. IV.

8. *Big Bang Cosmology: The Evolution of the Universe*

Nonmathematical introductions to cosmology include *Great Ideas and Theories of Modern Cosmology* (New York: Dover Publications, 1961) by J. Singh, and *The Structure and Evolution of the Universe* (New York: Harper & Bros., 1959). Herman Bondi's *Cosmology* (Cambridge: Cambridge University Press, 1961) is a more technical classic introduction, while "Cosmological Models and Their Observational Validation" by W. Davidson and J. V. Narlikar, in *Reports on Progress in Physics* 29 (1966), gives an excellent detailed discussion of more recent evidence. The controversial nature of the subject has not diminished since this was written. A good astronomical introduction which ties Essay 6 to this one is *Galaxies and Cosmology* by P. W. Hodge (New York: McGraw-Hill, 1966).

9. *A Quantum Universe: The Beginning of Time*

The nature of theoretical science is succinctly and clearly explained in Richard P. Feynman's *The Character of Physical Law* (London: British Broadcasting Corporation, 1965). Of historical interest is Florian Cajori's presentation of Isaac Newton's theories of mechanics and gravitation, *Mathematical Principles of Natural Philosophy* (Berkeley: University of California Press, 1934). Max Born's little paperback book, *Einstein's Theory of Relativity* (New York: Dover Publications, 1962) is perhaps the best introduction to relativity theory for the discriminating layman.

Informative popular accounts, though slightly dated, of our Universe and its origin are *The Universe*, a Scientific American Book (New York: Simon & Schuster, 1957), and George Gamow's *The Creation of the Universe* (New York: Viking Press, 1952). A solid introduction to the quantum theory is found in *Quantum Physics* by Eyvind H. Wichmann (Vol. 4 of the Berkeley Physics Course; New York: McGraw-Hill, 1971) and in Volume III (be sure to look at the other two volumes!) of R. P. Feynman's *Lectures On Physics* (with Robert B. Leighton and Matthew Sands; New York: Addison-Wesley, 1965).

Three semipopular articles in the periodical *Physics Today* clarify and update the topics covered in these books. An integrated view of modern relativity is provided by Mendel Sachs in "Space, Time and

Elementary Interactions in Relativity" 22: 51 (February 1969); the early moments of big-bang cosmology are well portrayed by Edward R. Harrison's "The Early Universe" 21: 31 (June 1968); and an interesting interpretation of the quantum theory, with possible cosmological consequences, is given by Bryce S. DeWitt in "Quantum Mechanics and Reality" 23: 30 (September 1970).

By far the best explanation of superspace, quantum gravitation, and quantum geometrical foam is the small book (very readable, but in German) by John A. Wheeler, *Einstein's Vision* (New York: Springer-Verlag, 1968). These topics are accessible to the general reader, though the going may become hard at times, in the following works: J. A. Wheeler, *Geometrodynamics* (New York: Academic Press, 1962); J. A. Wheeler, "Geometrodynamics and the Issue of the Final State," a chapter in *Relativity, Groups and Topology*, ed. C. DeWitt and B. DeWitt (New York: Gordon & Breach, 1964); B. K. Harrison, K. Thorne, M. Wakano, and J. A. Wheeler, *Gravitation Theory and Gravitational Collapse* (Chicago: Chicago University Press, 1965); and J. A. Wheeler, "Our Universe: The Known and the Unknown," *American Scientist* 56: 1 (1968). Of these four references, the first two are of intermediate difficulty, the third penetrates difficult ground, and the fourth is introductory.

Armand Wyler's work is cited in a popular article in *Physics Today* 24: 17 (August 1971).

10. Was There Really a Big Bang?

To persuade the reader that "big bang" does not necessarily subsume all matters cosmological, the following extensive list of references is cited in this essay:

R. A. Alpher, *Physical Review* 74: 1577 (1948).

R. A. Alpher and R. C. Herman, *Physical Review* 74: 1737 (1948).

——, *ibid.*, 75: 1089 (1949).

——, *Reviews of Modern Physics* 22: 153 (1950).

V. A. Ambartsumian, *Solvay Conference on Structure and Evolution of the Universe*, ed. R. Stoops (Brussels: 1958), p. 241.

——, *Structure and Evolution of Galaxies*, Proceedings of the 13th Conference on Physics, University of Brussels (New York: John Wiley and Sons, 1965), p. 1.

W. A. Baum, *Astronomical Journal* 62: 6 (1957).

——, *I.A.U. Symposium No. 15*, ed. G. C. McVittie (New York: Macmillan, 1961), p. 390.

A. S. Bennett, *Memoirs of the Royal Astronomical Society* 68: 163 (1962).

J. G. Bolton, *Astronomical Journal* 74: 131 (1969).

H. Bondi and T. Gold, *Monthly Notices of the Royal Astronomical Society* 108: 252 (1948).

H. Bondi, T. Gold, and D. Sciama, *Astrophysical Journal* 120: 597 (1954).

W. B. Bonnor, *Zeitschrift für Astrophysik* 39: 143 (1956).

K. Brecher, G. Burbidge, and P. Strittmatter, *Comments on Astrophysics and Space Physics*, in press.

E. M. Burbidge, G. R. Burbidge, W. A. Fowler, and F. Hoyle, *Reviews of Modern Physics* 29: 547 (1957).

G. R. Burbidge, *Annual Reviews of Astronomy and Astrophysics* 8: 369 (1970).

G. A. Day, A. T. Shimmins, R. J. Ekers, and D. J. Cole, *Australian Journal of Physics* 9: 35 (1966).

P. Demarque, J. G. Mengel, and M. Aizenman, *Astrophysical Journal* 163: 37 (1971).

R. H. Dicke, *Astrophysical Journal* 152: 1 (1968).

H. Dingle, *Observatory* 73: 42 (1953).

W. A. Fowler, *Rutherford Symposium Volume* (1968).

——, *Comments on Astrophysics and Space Physics* 2: 134 (1970).

A. Friedmann, *Zeitschrift für Physik* 10: 377 (1922).

G. Gamow, *Physical Review* 70: 572 (1946).

——, *ibid.*, 74: 505 (1948).

——, *Kgl. Danske Videnskab Selskab Mat. Fys. Medd.* 27, no. 10 (1953).

E. R. Harrison, *Monthly Notices of the Royal Astronomical Society* 148: 119 (1970).

S. W. Hawking and R. J. Tayler, *Nature* 209: 1278 (1966).

J. R. Houck and M. Harwit, *Astrophysical Journal Letters* 157: L 45 (1969).

F. Hoyle, *Monthly Notices of the Royal Astronomical Society* 108: 372 (1948).

——, *Proceedings of the Royal Society* A 308: 1 (1969).

F. Hoyle and G. R. Burbidge, *Nature* 210: 1346 (1966).

——, *ibid.*, 227: 359 (1970).

F. Hoyle and J. V. Narlikar, *Proceedings of the Royal Society* A 290: 177 (1966).

E. P. Hubble, *Proceedings of the National Academy of Science* 15: 168 (1929).

———, *The Realm of the Nebulae* (New York: Oxford University Press, 1936).

M. L. Humason, *Proceedings of the National Academy of Science* 15: 167 (1929).

M. L. Humason, N. U. Mayall, and A. R. Sandage, *Astronomical Journal* 61: 97 (1956).

I. Iben and J. Faulkner, *Astrophysical Journal* 153: 101 (1968).

I. Iben and R. Rood, *Astrophysical Journal* 161:587 (1970).

J. Kristian and A. R. Sandage, *Astrophysical Journal* 162: 391 (1970).

G. Lemaître, *Ann. Soc. Sci. Bruxelles* 47: A 49 (1927).

———, *Monthly Notices of the Royal Astronomical Society* 91: 483 (1931).

E. Lifshitz, *Journal of Physics* (U.S.S.R.) 10: 116 (1946).

E. Lifshitz and I. M. Khalatnikov, *Advances in Physics* 12: 185 (1963).

M. Longair, *Monthly Notices of the Royal Astronomical Society* 133: 421 (1966).

M. Longair and P. A. G. Scheuer, *Nature* 215: 919 (1967).

B. Y. Mills, *Paris Symposium on Radio Astronomy*, ed. R. N. Bracewell (Stanford: 1959).

R. Minkowski, *Astrophysical Journal* 132: 908 (1960).

D. Muelhner and R. Weiss, *Physical Review Letters* 24: 742 (1970).

J. D. North, *The Measure of the Universe* (Oxford: Clarendon Press, 1965).

I. D. Novikov and Y. B. Zeldovich, *Annual Reviews of Astronomy and Astrophysics* 5: 627 (1967).

J. B. Oke and A. R. Sandage, *Astrophysical Journal* 154: 21 (1968).

L. M. Ozernoy and A. D. Chernin, *Soviet Astronomy* 11: 907 (1967).

———, *ibid.*, 12: 901 (1968).

J. V. Peach, *Astrophysical Journal* 159: 753 (1970).

———, paper presented at I.A.U. Symposium No. 44, Uppsala, Sweden (1970).

A. Penzias and R. Wilson, *Astrophysical Journal* 142: 419 (1965).

V. Petrosian, E. E. Salpeter, and P. Szekeres, *Astrophysical Journal* 147: 1222 (1967).

M. Ryle, *Observatory* 75: 137 (1955).

———, *Annual Reviews of Astronomy and Astrophysics* 6: 249 (1968).

M. Ryle and P. A. G. Scheuer, *Proceedings of the Royal Society* A 230: 448 (1955).

M. Ryle and P. F. Scott, *Monthly Notices of the Royal Astronomical Society* 122: 389 (1961).

A. R. Sandage, *Astrophysical Journal* 127: 513 (1958).

——, *ibid.*, 133: 355 (1961).

——, *Proceedings of "Enrico Fermi" Physics Course 35* (New York: Academic Press, 1966), p. 10.

——, *Observatory* 92: 91 (1968).

——, *Physics Today* 23: 34 (1970).

——, paper presented at Semaine d'Etude "Les Noyaux des Galaxies," Vatican (April 1970), in press.

——, paper presented at Mayall Symposium, Tucson, Arizona (May 1971).

M. Schmidt, *Astrophysical Journal* 151: 393 (1968).

——, *ibid.*, 162: 371 (1970).

A. T. Shimmins, J. Bolton, and J. V. Wall, *Nature* 217: 818 (1968).

K. Shivanandan, J. R. Houck, and M. Harwit, *Physical Review Letters* 21: 1460 (1968).

I. S. Shklovsky, *Astrophysical Journal Letters* 150: L 1 (1967).

J. Stebbins and A. E. Whitford, *Astrophysical Journal* 108: 413 (1948).

K. Thorne, *Astrophysical Journal* 148: 51 (1967).

R. Wagoner, W. A. Fowler, and F. Hoyle, *Astrophysical Journal* 148: 3 (1967).

A. E. Whitford, *Astrophysical Journal* 120: 599 (1954).

——, *Astronomical Journal* 61: 353 (1956).

Index

Amino acids, 68, 83, 84–85
Atoms:
  electron binding, 73, 74–76
  nuclei, 73, 76–78
  radioactivity, 17, 73, 77–78

Big-bang universe, 10, 129–30, 143–52
  age, 147
  anisotropic models, 160–61
  causal horizon, 162–63
  evolution, 153–54, 162–63
  helium, 101, 147, 149, 177–79
  historical background, 166–67
  Newtonian models, 148
  source counts, 171–76
Black hole, 23–24

Cheshire cat, 151
Color-magnitude plot, 15, 16
Continental drift, 144–45
Cosmic nucleosynthesis, 101, 149, 153, 177–79
Cosmic rays, 36, 39
Crab nebula, 23, 35–43
  Chinese guest star, 35
  prototype of violent events, 42–43
  radiations, 36–37
  wisps and filaments, 38, 39, 41
  X-ray source, 36, 41

"Dark" companions, 29–30, 31
Darkness of night sky, 107–10, 127
Deceleration parameter, 146, 168–71
Doppler effect (red shift), 7, 91, 102–3, 105, 117–20, 146, 169
Dwarf star:
  black, 22
  red, 22
  white, 22–23

Elementary particles, 71, 74, 149, 157–58, 162–64
Exclusion principle, 21, 74–75, 80, 158

Galaxies, 6, 90–106
  angular momentum, 104
  chemical composition, 100–101
  classes, 93–97, 101–4, 169
  clusters of, 103–4, 109–10, 168
  colors, 96, 97–98, 171
  definition, 92–93
  determining red shifts, 169
  evolution, 99–100, 102, 104, 145, 146, 149–50, 170, 179
  explosions in nuclei, 104–6
  formation, 10
  helium content, 101
  hydrogen content, 93, 97, 98–99, 100
  masses, 95, 103
  Milky Way, 90–91, 103
  selection effects, 95–96
Gas:
  degenerate, 21–22
  perfect, 17–18
General relativity, 10–11, 148, 155–56, 158
Genetic code, 84–85, 144
Gravitational radiation, 11, 39, 157, 162–63

Herschel, John, 7–8
Hertzsprung-Russell diagram, 14–15
  main sequence, 15, 19–20
Hubble expansion, 6–7, 91, 92, 102, 108, 114, 116, 125, 127, 146–47, 166, 168–70

Life:
  DNA, 79, 84–85
  enzymes, 86, 87

Life (*cont.*)
    necessity of chemical forces, 80
    nonequilibrium environment, 72, 82,
        87–88
    precursors in molecular clouds, 68,
        83
    "primeval soup" origin, 83, 88
    probability of, 33–34, 89
    requirements (tradition, food, sun-
        shine), 83–88
    suitable locations, 88–89
Light:
    aberration effect, 117, 120–23
    corpuscular theory, 114–15
    inverse-square law, 110–11
    wave theories, 114–15

Maser effect, 57
Matter:
    elementary particles, 71, 157–58
    interactions, 72–73
    microscopic structure, 78–80
    stability of, 4, 143–45
Metaphysical dilemma, 164
Microwave background radiation, 9–10,
        58–59, 147, 149, 167, 176–77
Molecular clouds:
    ammonia and water, 55–57
    criteria for molecular detection, 60,
        69–70
    density, 56
    dust grains, 65, 69, 76, 82
    early observations, 54–55
    formaldehyde, 57–59
    "holes in the sky," 54, 58
    interstellar chemistry, 62
    inverse masers, 58–59
    "life soup" (bioclouds), 68
    masers, 57
    model of, 65–68
    molecules discovered, 61–64, 82

Neutron star, 23, 37, 38–39

Olbers paradox, 107–30

Planetary system:
    definition, 26–27
    formation, 30–31
    stability, 27–28

Pulsar, 23
    in Crab nebula, 37
    "glitches" (starquakes), 40
    unknown radiation mechanism, 42

Quantum gravitation, 155–58, 159–60
    "quantum foam" (fluctuations), 161–
        62
Quantum mechanics, 156
Quasars:
    discovery, 8
    properties: 9, 42–43, 102, 105–6, 145–
        46
    spatial distribution, 145, 173–74

Radio source counts, 171–75
    selection effects, 172
Red shift, *see* Doppler effect

Scalar-tensor gravitation, 148, 166
Scientific progress, 165
Solar system, 4–5, 25–26, 145
Special relativity, 127–28, 158
    density transformation, 123–25
    length contraction, 123–25
Stars:
    angular momentum (rotation), 27, 32
        39
    Cepheid variables, 91–92
    clusters of, 13–16
    double (binary), 33
    equilibrium, 17
    evolution, 20, 38–39, 81
    formation, 19, 30–31, 145, 150
    interior conditions, 18–20
    Kelvin-Helmholtz contraction, 16, 18–
        19, 30
    low-mass, 29–30
    magnetic fields, 13, 39, 40
    thermonuclear fusion, 17
Steady-state universe, 10, 114, 148–49,
        166–67
    creation of matter, 115–16, 163–64,
        167
    Olbers paradox in, 129–30
    source counts, 171–76
Stebbins-Whitford effect, 171
Sun, 13–20
    formation of, 150

Sun (*cont.*)
    radiation from, 82, 109
    solar wind, 13
    sunspots, 13
Supernova, 23, 35
    remnants, 36, 41
Superspace, 158–62
Synchrotron radiation, 36, 40–41, 105

Time, 154–55
Titius-Bode relation, 25–26, 27

Uncertainty principle, 21, 74–75, 79, 80, 157
Universe:
    contemplation of, 12
    expansion of, 7
    see also Big-bang universe *and* Steady-state universe